BIOAESTHETICS

posthumanities

CARY WOLFE, SERIES EDITOR

43 Bioaesthetics: Making Sense of Life in Science and the Arts
Carsten Strathausen

42 Creaturely Love: How Desire Makes Us More and Less Than Human
Dominic Pettman

41 Matters of Care: Speculative Ethics in More Than Human Worlds
Maria Puig de la Bellacasa

40 Of Sheep, Oranges, and Yeast: A Multispecies Impression
Julian Yates

39 Fuel: A Speculative Dictionary
Karen Pinkus

38 What Would Animals Say If We Asked the Right Questions?
Vinciane Despret

37 Manifestly Haraway
Donna J. Haraway

36 Neofinalism
Raymond Ruyer

35 Inanimation: Theories of Inorganic Life
David Wills

34 All Thoughts Are Equal: Laruelle and Nonhuman Philosophy
John Ó Maoilearca

33 Necromedia
Marcel O'Gorman

32 The Intellective Space: Thinking beyond Cognition
Laurent Dubreuil

31 Laruelle: Against the Digital
Alexander R. Galloway

30 The Universe of Things: On Speculative Realism
Steven Shaviro

29 Neocybernetics and Narrative
Bruce Clarke

28 Cinders
Jacques Derrida

27 Hyperobjects: Philosophy and Ecology after the End of the World
Timothy Morton

26 Humanesis: Sound and Technological Posthumanism
David Cecchetto

25 Artist Animal
Steve Baker

24 Without Offending Humans: A Critique of Animal Rights
Élisabeth de Fontenay

(continued on page 306)

BIOAESTHETICS

Making Sense of Life in Science and the Arts

CARSTEN STRATHAUSEN

posthumanities 43

University of Minnesota Press
Minneapolis
London

Published by the University of Minnesota Press
111 Third Avenue South, Suite 290
Minneapolis, MN 55401-2520
http://www.upress.umn.edu

Library of Congress Cataloging-in-Publication Data
Names: Strathausen, Carsten, author.
Title: Bioaesthetics : making sense of life in science and the arts / Carsten Strathausen.
Description: Minneapolis : University of Minnesota Press, 2017. | Series: Posthumanities ; 43 | Includes bibliographical references and index. |
Identifiers: LCCN 2017005483 (print) | ISBN 978-1-5179-0074-8 (hc) | ISBN 978-1-5179-0075-5 (pb)
Subjects: LCSH: Science and the arts. | Biology. | Aesthetics.
Classification: LCC NX180.S3 S77 2017 (print) | DDC 701/.05—dc23
LC record available at https://lccn.loc.gov/2017005483

UMP LSI

Und wir: Zuschauer, immer, überall,
dem allen zugewandt und nie hinaus!
Uns überfüllts. Wir ordnens. Es zerfällt.
Wir ordnens wieder und zerfallen selbst.

And we: onlookers, always, everywhere,
always looking into, never out of, everything.
It fills us. We arrange it. It collapses.
We arrange it again, and collapse ourselves.

RAINER MARIA RILKE

CONTENTS

ACKNOWLEDGMENTS

This book started some ten years ago as a project on media history and aesthetics. It was supported by a generous eighteen-month Humboldt Fellowship from 2006 to 2007, which I spent both at the Zentrum für Literatur- und Kulturforschung (ZfL) in Berlin and at the Zentrum für Kunst und Medientechnology (ZKM) in Karlsruhe. I particularly wish to thank Dr. Peter Waibel for his mentorship and support during my visits to Karlsruhe. I also want to express my gratitude to the many friends and colleagues who have read and commented on earlier drafts of this manuscript, in particular Marco Abel, André Arief, Roger Cook, Stefani Engelstein, Sean Franzel, Peter Gilgen, Noah Heringman, Sean Ireton, Ted Koditschek, Kristin Kopp, Brad Prager, Philipp Robbins, Roland Vegso, and Jake Young. I owe special thanks to Chris Breu and Cary Wolfe, who provided expert support and insight at crucial times during the completion of this project. The book is dedicated to my wife, Jennah; our son, Coen; and my daughter, Clara; and my friends and colleagues at the University of Missouri.

INTRODUCTION

> Objectivity means that you look very hard at those things which you choose to look at.
>
> **GREGORY BATESON**

The Biological Nature of *What*?

In 2000, Ian Hacking, the renowned philosopher and historian of science, published a book titled *The Social Construction of What?* Written during the science wars of the 1990s and published a few years after the infamous Sokal hoax in 1996, Hacking's study provided a trenchant critique of the various constructivist theories that influenced academic discourse at a time when scholars in all fields sought to expose the constructed nature of ever more phenomena. Although polemical at times, the main goal of the book was less to expose the fashion-driven nature of academic discourse than to clarify the distinct—and at times even mutually exclusive—meanings of the term "constructivism" when applied to specific social contexts studied by different disciplines. Words like "social construction," Hacking cautioned, "can work like cancerous cells. Once seeded, they replicate out of hand."[1]

The question in my subhead title—"The Biological Nature of *What*?"—is meant to indicate that similar mutations characterize the rapid proliferation of biological concepts (such as "evolution," "gene," and "human nature") in the humanities over the past few decades. There can be little doubt that today, less than twenty years after the publication of Hacking's book, the prevailing academic mood has swung to the opposite side of the spectrum. Scholars' erstwhile obsession with social constructivism has given way to our current fascination with what might be called "biologism"—the effort to understand all aspects of human culture, including art and politics, in biological terms as part of our evolutionary heritage. As evidence of this trend, one need only look at the number of scholarly publications and nascent academic fields that carry the prefix "bio" in their title. Examples include "biophilosophy," "biopolitics," "bio-art," "bioethics," "biopoetics," "biotechnology," "biomedia," "bioeconomics," and "biocapital," to name but the most prominent ones.[2]

This ongoing proliferation of "bio" discourses clearly testifies to the dominant role of biology in both scholarly and popular literature today. Similar to early twentieth-century physics whose place it usurps, biology is widely considered the new *Leitwissenschaft* of the twenty-first century. The crucial goal for college administrators today, according to business professor and innovation guru Clayton M. Christensen, must be to "change the DNA of higher education from the inside out."[3] The growing influence of the life sciences is particularly evident in contemporary aesthetic theory. Joseph Carroll, the founder of "literary Darwinism," is one among many who proclaim a "Darwinian revolution in the humanities."[4] Carroll's view is echoed by numerous scholars from a wide variety of disciplines that range from neurophysiology and cognitive theory to art, literature, and film studies: "Only by knowing where art comes from biologically will we know what it is and what it means," the anthropologist Ellen Dissanayake insists.[5] Semir Zeki, an accomplished scholar of neurobiology who was named the first professor of neuroesthetics at the University College London in 2008, agrees: "No theory of aesthetics which does not have a strong biological foundation is likely to be complete, let alone profound."[6]

This premise is hardly new, of course. "All aesthetic values are based on biological values," Nietzsche already proclaimed at the end of the nineteenth century, adding that all "aesthetic pleasures are biological pleasures."[7] In fact, from its very beginning in the eighteenth century, when Alexander Gottlieb Baumgarten first defined aesthetics as a "science of sensory knowledge," the history of Western aesthetics has struggled with the body–mind dualism and the tug-of-war between speculative-idealist and scientific-materialist views on art and human culture. The rise of positivism in the second half of the nineteenth century, for example, resulted in numerous physiological studies on art and human perception both in Germany (e.g., Wilhelm Wundt, Hermann v. Helmholtz, Gustav Fechner) and abroad (e.g., William James). In contrast to German idealism and the romantic tradition, the scientific strand of aesthetic theory emphasized the embodied quality of human experience, focusing in particular on the basic emotions and affects that exist below the threshold of human consciousness.

Overall, however, there is no denying that the modern history of aesthetics, from Hegel to Heidegger, has been dominated by what Jean-Marie Schaeffer calls "the speculative theory of art."[8] The speculative tradition, Schaeffer convincingly argues in his comprehensive study, "projects a transcendent entity that is supposed to *found* the diversity of artistic practices and to have ontological priority over them."[9] Concerned with the putative essence and function of art, the speculative theory is, above all, normative. It emphasizes the "truth-content" of art so as to distinguish art from artifact, beauty from kitsch. Of course different philosophers have had different ideas about how to define and access the truth of art. But they nonetheless agreed that this truth exists and that it informs both the production and the reception of art. If there is "truth" apart from and outside of the rational-scientific view of the modern world, according to the speculative theory, it resides within "true" works of art.

The current rise of biologism across the humanities is primarily a reaction against this long-standing speculative tradition of art and aesthetics in Western culture. What distinguishes this reaction from its historical predecessors in the nineteenth century, however, are its broad universalist claims buttressed by recent sociobiology and neo-Darwinian theory. The prevailing sense among today's scholars interested in biology and culture is that evolutionary theory will not merely supplement, but help revolutionize the traditional modes of inquiry in the humanities. "To adopt a human nature (or 'adaptationist') point of view," Dissanayake declares, "is a revolution in worldview for the humanities that can be likened to replacing a geocentric with a heliocentric perspective in cosmology."[10] The condescending tone of this statement betrays the strong scientistic bias that underlies and unites biologistic theories of culture. The humanities, according to Dissanayake's far-fetched analogy, will remain trapped in the dark ages of ideological superstition until they finally learn to adopt the correct Darwinian perspective of how things really are.

This view is widely shared among social scientists and increasingly prominent among literary scholars and film critics as well. Next to traditional aesthetics, several other new research fields and disciplines have recently emerged in the humanities (such as "literary Darwinism," "neuroaesthetics," and "cognitive cultural studies") that tend to embrace a—frequently distorted—version of Darwinian evolutionary theory as the harbinger of a new research paradigm in the humanities at large.[11] "If evolution took place," the Slavicist Brett Cooke, a pioneer in the field of biopoetics points out, "then all of our prior perspectives on the arts, if not on the nature of perception itself, will have to be reconsidered in our new evolutionary paradigm."[12] Joseph Carroll, the founder and leading scholar of literary Darwinism, agrees: "At least among their most ambitious adherents, they [literary Darwinists] aim not at being just one more 'school' or 'approach.' They aim at fundamentally altering the paradigm within which literary study is now conducted."[13]

Apart from the current rise of biologism, the prominence of the digital humanities and posthumanities also suggests that the speculative theory of art has run its course. Rather than deconstruction or traditional hermeneutics, more and more humanities scholars today advocate the use of empirical models for the study of art and culture.[14] This methodological shift seems to indicate that the humanities increasingly suffer from what might be called "hermeneutic fatigue syndrome"—and understandably so. After a century of self-reflexive drilling into the being of language, younger scholars in particular are no longer swayed by the (post)modern "fetishization of language."[15] For while the speculative tradition remains stuck inside the hermeneutic circle, the empirical sciences continue to flourish outside it. The current popularity of Deleuzian philosophy across the humanities, I would argue, is due precisely to his anti-hermeneutic premise. It is well known that Deleuze denounced traditional "interpretations" of art as banal forms of "common sense," a notion he held in considerable low esteem.[16] Unimpressed by the linguistic turn, Deleuze remained skeptical of Saussurian semiotics throughout his life, which explains why he joined neither the structuralist nor the post-structuralist movement in the second half of the twentieth century. He strongly favored Peirce's semiology instead, which, unlike semiotics, also includes other, nonlinguistic systems of signification. Many media and cultural critics have done the same. The critical self-reflexivity championed by traditional hermeneutics and deconstruction, they claim, cannot but end up reducing feelings to signs, sensations to propositions, and affects to thought. Speculative inquiry of this kind transforms the human body into a "discursive body," one that "performs" rather than simply feels, moves, and senses.[17]

To a large degree, hermeneutics itself is to blame for this backlash, because it remained overly focused on the medium of language as the sine qua non of thought. Traditional hermeneutics failed to acknowledge that meaning takes shape not only in language, but in numerous other, nonlinguistic media and communication systems as well, including images, graphs, numbers, computers, human brains, and so forth. Similar to Heidegger and Derrida, Hans-Georg Gadamer, too, exalted language as the "middle ground *[die Mitte]*" where all human thought and understanding takes place.[18] Hermeneutics, for Gadamer, "bring[s] to light a *pre-ordered* dimension that *precedes* all application of scientific methods."[19] Which is to say that Gadamer claimed the same epistemological supremacy for hermeneutics that neo-Darwinians claim for evolutionary theory today. Much like biologism tries to absorb aesthetics into evolutionary theory, Gadamer once insisted that "*aesthetics has to be absorbed into hermeneutics.*"[20] Gadamer's insistence on the supreme—as opposed to distinctive—epistemological power of hermeneutics is certainly one reason why humanities scholars have grown tired of the speculative art of interpretation and begun to embrace biologism instead.

But there are also important institutional reasons behind the prominence and revolutionary rhetoric of today's biologism. The erosion of public financing in higher education over the last several decades has led to an enormous increase in private investments focused on the natural sciences as opposed to the humanities.[21] Most scholarly advances in the areas of molecular biology, genetics, and neuroscience—including the development of advanced diagnostic tools such as fMRIs and PET scans—would have been impossible to achieve without significant corporate sponsorship and other private support for scientific research. While the main objective of such investments is to ensure that the academic focus remains on the "applied" side of science in order to yield profitable tools and services for the private market, the scarcity of public funding also creates significant pressure on other, non-scientific disciplines to keep pace and clarify their own contribution to the study of "life"—yet another concept increasingly popular among scholars from all disciplines.

In a time of intense competition for funding and resources, it might seem like a sensible approach for the humanities not only to seek an institutional alliance with the sciences, but to help unite all human knowledge under a single methodological umbrella, as E. O. Wilson famously argued in his book *Consilience* from 1998: "The central idea of the consilience world view is that all tangible phenomena, from the birth of the stars to the working of social institutions, are based on material processes that are ultimately reducible, however long and tortuous the sequences, to the laws of physics."[22] This book, by contrast, argues that this putative consilience between the disciplines ultimately leads to an impoverished understanding of human culture and its history. Much as bioaesthetics embraces scientific insights, it keeps its distance from the "universal acid" spilled by neo-Darwinian theory.[23] To some degree, my argument overlaps with that of Stephen Jay Gould, the well-known paleontologist and evolutionary theorist. In his last book, with the playful title *The Hedgehog, the Fox, and the Magister's Pox: Mending the Gap between Science and the Humanities*—which serves to emphasize the often unacknowledged power of metaphorical language in scientific research—Gould provided a thorough critique of Wilson's concept of consilience and clarified the conceptual and methodological problems at stake. His comments are worth quoting at length:

> I use the fox and the hedgehog as my model for how the sciences and humanities should interact because I believe that neither pure strategy can work, but that a fruitful union of these seemingly polar opposites can, with goodwill and significant self-restraint on both sides, be conjoined into a diverse but common enterprise of unity and power. The way of the hedgehog cannot suffice because the sciences and humanities, by the basic logics of their disparate enterprises, do different things, each equally essential to human wholeness. We need this wholeness above all, but cannot achieve the goal by shearing off the legitimate differences that make our lives so varied, so irreducibly, and so fascinatingly, complex. But if we lose sight of the one overarching goal—the hedgehog's insight—underneath the legitimately different concerns and approaches of these two great ways, then we are truly defeated, and the dogs of war will disembowel our underbellies and win.[24]

Gould aptly emphasizes the different "basic logics" that distinguish "the legitimately different concerns and approaches" of heterogeneous disciplines. This part of Gould's argument I fully agree with: the sciences and the humanities are fundamentally irreducible to one another.[25] These differences, in my view, are most apparent in the different media used to present academic research. At professional gatherings, the sciences prefer poster sessions and PowerPoint presentations that include empirical data and mathematical formulae. The humanities, on the other hand, continue to rely almost exclusively on the traditional format of oral lectures based on written manuscripts. Each method, for sure, has its advantages and disadvantages, yet the role of language in each context is fundamentally different: for the sciences, everyday language is just one of many systems of communication like numbers and images, charts and graphs; whereas for the humanities, language is clearly a privileged medium to make the point they want to make. In the social sciences, for example, the going practice is simply to refer (in parenthesis) to a particular author and publication without actually quoting the text in question. The underlying assumption seems to be that the specific wording in the original is either irrelevant or at best secondary to the supposedly self-evident ideas intended by the author and summarized by the critic. It is as if these writers assumed that everybody who read, as an example, *The Phenomenology of Spirit* more or less took away the same message; that there is no semantic ambiguity in Hegel's text; and that his ideas must be synthesized on the argumentative level, not analyzed on the linguistic level.

There is nothing inherently wrong with this assumption, given the nature of scientific research and its focus on empirical data rather than semantic propositions. Yet it seems equally obvious that such a practice, from the humanities' perspective, takes for granted what remains to be proven—namely, the specific meaning of a sentence or paragraph. By and large, the sciences are content to treat language as a transparent medium for the representation of abstract ideas, whereas the humanities assume that these ideas are fundamentally shaped by the very language used to express them. That language needs to be respected and the ideas examined as they are formulated in the text. The right way to economize in a scholarly monograph is to be sparse with one's own words, not that of one's authors. To summarize their ideas without letting their own voices be heard is to meddle with the scene of the investigation and taint its object of study. The library, after all, is the humanists' laboratory, and the overall quality of our research is largely determined by the accurate interpretation of the texts we consult. Fatigued or not, the practice of close readings remains important for political reasons as well. In a time of Twitter logic and "alternate facts" that increasingly resembles a 3-D reality show of George Orwell's *Nineteen Eighty-Four*, it is absolutely crucial for democratic institutions like academia to remain truthful to historical facts and to raise awareness about the various media and levels of representation that constitute the public sphere and contribute to the production of knowledge. These media include not only the rhetorical use of language, but also the use of technological devices and the myriad ways politicians (mis)represent scientific research to (mis)inform the public.

I also agree with Gould's second point that the different methodologies of science and humanities are joined together by the common purpose of producing knowledge (what Gould calls their "common enterprise of unity and power"). However, I do not think we should interpret this "common enterprise" to mean that "the one overarching goal" of all academic inquiry is the pursuit and/or discovery of "human wholeness," as Gould suggests. The very idea of "human wholeness" is part of the problem, not the solution. Human beings are not whole, nor shall they ever be.[26] It is but a small step from Gould's notion of "human wholeness" to E. O. Wilson's claim that our shared "human nature" will soon be explicable in scientific terms alone: "As the properties of mind are clarified empirically," Wilson declared in 2005, "it will also be possible to define human nature with greater precision. That there is a human nature is no longer in doubt."[27] Once we assume that human wholeness exists, the ensuing debate will likely reflect the current prowess of the life sciences to define the terms and interpret the results. It will privilege the empirical and "objective" terms of modern science as opposed to the more speculative and socio-historical terms used in the humanities. This might explain why even the most accomplished scientists, like the Nobel laureate Eric Kandel, cling to the allegedly egalitarian goal of "unifying" diverse disciplines, such as neuroscience and art history, by "incorporating" one into another.[28]

Against Consilience

A central premise of bioaesthetics is that human nature is neither whole nor one, but multiple and fractured across time and space. We are not just defined by who we were in the past, but also by what we may become in the future. Hence there are good reasons for bioaesthetics to resist the idea of "human wholeness" or "human nature" altogether. For one, given its self-referential logic, the very attempt to objectify human nature gives rise to a *circulus vitiosus*, as Hannah Arendt (and other political philosophers) have argued:

> The problem of human nature seems unanswerable in both its individual psychological sense and its general philosophical sense. It is highly unlikely that we, who can know, determine, and define the natural essences of all things surrounding us, should ever be able to do the same for ourselves—this would be like jumping over our own shadows. . . . The question about the nature of man is no less a theological question than the question about the nature of God; both can be settled only within the framework of a divinely revealed answer.[29]

This is basically a Kantian position: the question about "human nature" is unanswerable, because it constitutes an antinomy of reason. Arendt's further point is that any reference to human nature is not just epistemologically unproductive, but politically dangerous, because it can easily be misappropriated to subordinate difference to sameness, human specificity to universality. "The human" does not exist, Arendt points out again and again; only humans do.

Another reason to resist the quest for human wholeness is that it leads bioaesthetics down a one-way street: scientific knowledge is exemplified with reference to art—be it a literary text (e.g., Joseph Carroll, Brian Boyd), a painting (e.g., Dennis Dutton, Jean-Pierre Changeux) or a piece of music (e.g., Oliver Sacks)—yet without contextualizing the particular object of study within its own historical, material, or discursive environment. Outside this contextualization, however, art falls silent. Once its specificity is dissolved within the universal acid of neo-Darwinian theory, aesthetic experience can no longer challenge the concepts or methods applied to it. Biologism calls on art for the sole purpose to verify what it already knows. It fails or refuses to acknowledge the fact that the concepts we use codetermine the objects we analyze. Consilience is doomed to fail, because there has never been—nor could there ever be—a neutral middle ground between the two cultures. The middle ground is always already occupied and conceived differently by each side. It is not just an empty conceptual space waiting to be defined in rational terms, but an emotional, visceral space beset by human habits, disciplinary boundaries, and personal beliefs. This, precisely, is what evolutionary theory has taught us about organisms and their environment.

And yet, the fantasy of this epistemological neutrality continues to dominate the interdisciplinary debate between the sciences and the humanities. Who, today, would dare *not* to profess allegiance to some neutral "middle path" that allegedly leads beyond the binaries of social constructivism versus biological evolution, of first-person versus third-person perspectives, of nature versus nurture? The rhetorical endorsement of this imaginary middle ground extends across all disciplinary fields today, from the natural and the social sciences to the arts and humanities. The explicit goal of Barkow, Tooby, and Cosmides's influential 1992 anthology *The Adapted Mind*, for example, was precisely "to supply the missing middle" between biological inheritance and sociocultural constructions of human nature.[30]

A closer look at this rhetoric, however, quickly reveals the semantic plasticity of this metaphorical "middle ground." For it turns out that both sides, neo-Darwinians and their critics, claim to inhabit this middle ground—while, at the same time, they vehemently criticize each other for disregarding it. Although Richard Lewontin, for example, is a fervent critic of evolutionary psychology and *The Adapted Mind*, he, too, champions his own version of a "third way" beyond reductionism and constructivism.[31] As in all political debates, the other side is characterized as radical, monolithic and exclusionary, whereas one's own side is deemed moderate, flexible, and well balanced. Moreover, each side rhetorically positions itself as an intellectual minority desperately struggling against the academic mainstream and the disciplinary hegemony of their opponents' view. Thus, according to its proponents, neo-Darwinian theory provides a scientific *corrective* against social constructivism run rampant, whereas its critics regard it as evidence for the growing hegemony of the reductionist worldview in today's corporate university. Bioaesthetics, I fear, is no different in that regard. It, too, is biased and promotes what I hope will be a rigorous defense of the humanities to stand their intellectual and institutional ground against the increasing dominance of scientific approaches to the study of art and culture. When it comes to the two cultures, all academics are partisan.

As scientists and cultural critics trade barbs about what they perceive as an ideological—and potentially dangerous—denunciation of their methodology by the other side, one might reasonably assume that any "middle ground" between the two cultures "has remained unoccupied because of the perceived [sociopolitical] implications," as opposed to the "intellectual difficulty" of uniting evolution and constructivism, as David Sloan Wilson has suggested.[32] So, does the nature/nurture debate and the entire two-culture phenomenon simply boil down to politics after all? Would all our troubles be over if we could just get over the polemics and restart the conversation on the basis of "goodwill and significant self-restraint on both sides," as Stephen Jay Gould suggested in the passage quoted above?

The answer is no. Most of us—or at least those who enjoy the privilege of administrative duties or have participated in academic funding committees—should have encountered sufficient empirical evidence to conclude that academic questions are always and necessarily intertwined with political questions. Politics is never just about politics; it is always politics about something, and in the academy, some of our most pressing "political" questions are precisely methodological questions. They are questions about how best to study this or that—like art or human culture, for example. We may all agree in principle about the need for a neutral middle ground to help adjudicate these questions. A genuine consensus about the specific nature of this ground, however, can only be forged politically, not rationally. For who shall decide on the specific parameters and conceptual terrain necessary to define it? The political rhetoric between two fundamentally conflicting paradigms, I argue, is a *symptom,* not the *cause* of the failed dialogue between the two cultures. Academic disciplines are separated by real terminological, methodological, and epistemological differences between scientific studies and hermeneutic inquiry. The epistemological relation between these two paradigms cannot be determined "objectively" or "rationally," because concepts like reason and objectivity are themselves subject to the debate. Not even the seemingly neutral terrain of art and aesthetics—the area that, at least since Kant, allegedly bridges the divide between theoretical and practical reason, between knowledge and values—has been able to provide the magical middle ground both parties claim to seek.

In short, there is no middle ground as such, just as there is no biological "environment" or social "milieu" as such.[33] These terms are inherently relational; they designate concepts that take shape within particular contexts. The (concept of) middle space only exists in and through the discriminatory perspectives enabled by different disciplines, and there can never be an epistemologically "neutral" zone between them. Hence it makes little sense to pursue the consilience of knowledge, as Wilson suggested, unless all parties were to agree up front on a suitable metadiscourse of interdisciplinary exchange. Given its political nature, such a meta-discourse would have to respect the intellectual and institutional autonomy of the humanities' approach to, and understanding of, human culture and cognition. It would also have to acknowledge and reflect on the epistemological constraints of scientific theories, including Darwinian evolution. From the humanities' perspective, biologism has fallen short on both counts. It serves to demonstrate that little is gained and much is lost in the radical pursuit of scientific reductionism. Some time ago, Louis Menand claimed that consilience is "a bargain with the devil."[34] I think he was right, except that the "devil"—or the "dogs of war," to use Gould's metaphor—is more likely to look like your local dean or provost who regrets to inform you that any vacant faculty lines in the humanities will have to be reallocated to support new initiatives in STEM research and the life sciences across campus. That is neither war nor evil, but it is still bad enough, in my view. The only way for bioaesthetics to stem the ongoing erosion of the humanities is to keep a critical distance from biologism and consilience alike.

The substitution of biologism for constructivism in current academic discourse amounts to little more than replacing one totalizing logic with another. Today's biologism fetishizes material objectivity and scientific reductionism in much the same way that constructivism fetishized abstract concepts and social conventions. Both "isms" pit matter against meaning, objects against concepts, when, instead, they should concentrate their efforts to examine how each of them codetermines and reproduces the other. If nothing else, the recognition of this codependence of concepts and objects remains an important legacy of the constructivist frenzy of the 1990s. It confirms what twentieth-century historians of science (from Ludwig Fleck and Gaston Bachelard to Georges Canguilhem, Hans-Jörg Rheinberger, and Lorraine Daston) have argued all along—namely, that concepts are not passive reflections of the primordial state of affairs, but active participants in the construction of human environments and our knowledge about them.[35] The same is true of the entire methodological apparatus—the *dispositive* in Barthes's or Foucault's sense—at the heart of science. "Methods are not mere catalysts that remain external to their object," Hans-Jörg Rheinberger insists; "they modify and also exhaust themselves in their own application."[36]

Since knowledge consists of binding an object to a concept, or matter to meaning, it is precisely the discrepancy between the two, rather than their identity, that enables scientific practice. The use of scientific concepts, George Canguilhem points out, is "at once both inevitable and artificial," because the scientist "strives to grasp a becoming whose meaning is never so clearly revealed to our understanding as when it disconcerts it."[37] Concepts clarify that matter and meaning, thing and thought, co-emerge as intertwined beings in the world. And since concepts have a history that defines their meaning, it follows that scientific objects, too, are profoundly historical in nature. Both concepts and objects, in other words, are site-specific and co-determined by the sociohistorical context in which they appear: "Things," Lorrain Daston clarifies, "are both socially constructed *and* real. That is, they depend crucially on the cultural resources at and in a given context . . . *and* they capture some aspect of the world; they work. But they are neither historically inevitable nor metaphysically true. Rather they are contingent to a certain time and place yet valid for certain purposes."[38]

In order to further elucidate this crucial point, let me briefly return to Hacking's critique of constructivism mentioned above, which I believe holds just as true for biologism today. Like Daston, Hacking readily admits "that something can be both real and a social construction" at the same time.[39] Scientific objects, he declares, "are new things in the world, brought into being in the course of human history." They are "artificial phenomena" and "created effects" produced by scientific practice.[40] According to Hacking, however, once these objects have thus come into being, and once they have been used and manipulated scientifically, they must be considered real and should be released into the wild. These freshly minted objects—however imperceptible they may be to the unaided human sensorium—now exist as autonomous entities independent of the scientific practice that originally "invented" or "discovered" them. What separates Hacking's view from Daston's is precisely this shift of emphasis from the conceptual to the objective side of the scientific process. Unlike Daston, Hacking is less interested in epistemological questions about the being of concepts than in ontological questions about the independent being of material reality. "What follows is about scientific realism, not about rationality," he emphasized more than thirty years ago,[41] a position he reiterated more recently in his book *Historical Ontology*: "My historical ontology is concerned with objects or their effects which do not exist in recognizable form until they are objects of scientific study. Daston describes her applied metaphysics as about the coming into being of objects of study—and not the coming into being of objects, period."[42] After each successful experiment, Hacking argues, scientists can safely dismiss their former "object of study"—such as photons or microbes or any other previously unrecognizable object that came to light only within the artificial environment created in the laboratory—back into the material world "of objects, period." Hacking's emphatic rhetoric literally makes the point of cutting the umbilical cord between concept and object, using his "period." as a syntactic wedge to sever theory from reality. Daston, on the other hand, insists on the structural coupling of epistemological statements and ontological facts. She refuses to talk about objects severed from the conceptual and scientific arrangement that originally enabled their "coming into being." Bioaesthetics follows Daston rather than Hacking, because it emphasizes the historically changing relation between scientific concepts and the objects to which they refer. This, however, is not to claim that Hacking's version of scientific realism is wrong. It only means that his strong emphasis on the objective existence of things, much like the constructivist denial of this objectivity, tends to foreclose the detailed analysis of how—and why—things *gain* objectivity within specific sociohistorical contexts in the first place.

Speaking of objectivity, this is another lesson to draw from the science wars of the 1990s: objectivity has a history like any other concept. As Lorraine Daston and Peter Galison convincingly demonstrate in their extensive study, objectivity only emerged as an epistemic paradigm during the course of the nineteenth century, promising to establish "a new form of unprejudiced, unthinking, blind sight" as the foundation of positivist science.[43] And far from being opposed to humanities research at the time, the triumph of objectivity largely coincided with the rise of hermeneutics as a rigorous scientific practice in its own right. Today, however, the concept of objectivity has largely become unproductive, because it gives rise to epistemological posturing and standoffs that stifle rather than foster interdisciplinary collaboration. The best way to deal with objectivity, indeed, might be to avoid the concept altogether, because it provides little help for trying to understand how human beings enact—or interact with—the environment in which they live, as Peter Godfrey-Smith suggests.[44] If, however, one were to insist on retaining the concept nonetheless, then among the many available options for its proper definition—objectivity defined as "impartiality," or as "rational consensus," or as "a subjective construct," or as plain "nonsense"[45]—the one that comes closest to my understanding is that of Gregory Bateson. Objectivity, Bateson wisely stated, "means that you look very hard at those things which you choose to look at."[46] This is the kind of objectivity bioaesthetics ought to pursue.

Throughout this book, the paradoxical intertwinement of concepts and objects as both artificial and real will take center stage. My overall goal is to demonstrate that each academic discipline approaches this paradox from a different perspective, none of which is reducible to any other. Different concepts yield different objects of study, and each concept will acquire new meanings and create new referents once it takes root in different contexts. The ensuing epistemological problems and muddled meanings are unavoidable. There is no scientific cure for the paradoxical relation between concept and object born of modern science and philosophy, because this paradox grows at its very root: "Logos is paradox."[47] This explains why key anthropological concepts such as "life" or "human nature" cannot be defined in other than contradictory terms. Western theories of life, Eugene Thacker demonstrates in his comprehensive study *After Life,* display "fundamental contradictions inherent in the concept of life, contradictions that are logically incoherent, and yet ontologically necessary."[48]

Bioaesthetics does not aim to overcome the constitutive paradox of Western thought, but aims to explain and conceptualize its persistence in different historical and disciplinary contexts. Since "concepts are words in their historical sites," as Hacking puts it, they remain subject to historical change like any other entity (except, perhaps, Plato's immortal ideas).[49] Following Jacques Monod and Richard Dawkins, we might even go further and suggest that concepts are akin to organisms in the sense that they, too, constantly change over time. By force of this analogy, it should then also be clear that not every conceptual mutation automatically triggers a new referent or material object of study. Instead, the various lexemes or alleles of a particular concept lie dormant until their difference makes a difference—that is, until someone calls on them to give meaning to or make sense of that which hitherto had been unknown.

But that is about as far as this metaphorical understanding of concepts as organisms will allow before leading us astray. As I will elaborate in my third chapter, concepts are neither memes nor genes. Their world is discursive, not physical. They are subject to hermeneutic principles, not scientific laws. The difference between the two has been elaborated by philosophers throughout the twentieth century. Following Gadamer and Rorty, the Italian philosopher Gianni Vattimo puts it this way: "Hermeneutics, as a philosophical theory, 'proves' its own validity only by appealing to a historical process of which it proposes a reconstruction that shows how 'choosing' hermeneutics—as opposed to positivism, for example—is preferable or more justified."[50] This justification is always partial and never complete. Hermeneutics, like life, always turns back on itself to reexamine its own genesis. This recursivity, Rorty points out, produces the kind of "holistic, antifoundationalist, pragmatist treatments of knowledge and meaning" that distinguishes the humanities from the sciences.[51] Hermeneutics, therefore, is not the name for a particular academic discipline, nor should it be misconstrued as just another "interpretative method" allegedly in need of scientific guidance and objective foundations.[52] Hermeneutics is no more a method of reading than science is a technique for measuring. Instead, science and hermeneutics each constitute a comprehensive epistemological paradigm trying to make sense of being human.

Bioaesthetics, for sure, relies on *this sense* of hermeneutics, but not the kind that defines art exclusively in speculative terms such as Heidegger's "aletheia" or Gadamer's "truth." At the same time, however, bioaesthetics must also resist the current trend to define art exclusively in scientific terms, which is just as metaphysical an approach as traditional hermeneutics ever was: instead of fetishizing language and aesthetic truth, biologism fetishizes genetics and neo-Darwinian theory. On this point, I fully agree with Gadamer: if the humanities "were to accept the method of the natural sciences as the only norm to guide their inquiry, [it] would amount to their self-annihilation."[53] This epistemological incommensurability, however, is neither an invitation for bioaesthetics to disregard the sciences altogether nor a call to try and usurp their current academic status as the ultimate foundation of human knowledge. That status and its foundation is a myth. Bioaesthetics is based on the premise that the search for hermeneutic "truth" is more or less analogous to that for scientific "objectivity": neither is primary, and both are inevitably flawed, albeit each in its own unique way. The two paradigms are often complementary, but they do not add up like two halves of a puzzle that make a whole. There is no whole.

What Is Bioaesthetics?

Bioaesthetics demonstrates what the humanities have to offer to the sciences: a historical hermeneutics of concepts, or what the Germans call a *Begriffsgeschichte*. This book provides numerous case studies of the historical transformation of scientific concepts and their impact in the human lifeworld. This history can only be studied retrospectively, in hindsight and from within a particular socioepistemological system of knowledge. "A directory of concepts," Karlheinz Barck points out, "cannot be determined *a priori*, but is based on heuristic models determined by the foreknowledge we apply to them."[54]

I doubt many scientists would disagree with this statement. As my previous comments indicate, an increasing number of researchers in the life sciences explicitly acknowledge the need to clarify the semantic shifts of key biological concepts as they migrate across disciplinary fields both within and beyond the life sciences.[55] In his detailed discussion of Pasteur's study of microbes in 1858, Bruno Latour aptly demonstrated this codependence of matter and meaning in scientific research. Pasteur's scientific papers and laboratory notes, Latour argued, were not just summary accounts of the empirical evidence previously established in the laboratory. His texts also interacted with and helped produce this very evidence in the first place. Scientific facts, in other words, do not just speak for themselves. They must also be *articulated* as independent, objective, and real entities. This claim has nothing to with deconstruction or postmodern relativism, as hard-nosed realists continue to insist. Instead, it adequately describes how signification and referentiality function in any meaning-producing discourse, including science.

It is a slippery slope, though. Right after Latour made a good case for the dual nature of Pasteur's microbes, he goes one step further and ruins it. The crucial point about microbes, he said, "is not just that they are both made *and* real. Rather, it is precisely *because* they have been artificially made up that they gain a complete autonomy from any sort of production, construction, or fabrication."[56] So, yes, microbes are both made and real, but of course they had to be *artificially made up first*; only then did they become real. We shall encounter this kind of linear logic again and again in different contexts and debates throughout this book. Depending on discipline and research interests, scholars will argue with one another over linear order and first principles, about what comes first and which paradigm beats the other: Does the autopoiesis of living systems precede their replication, or does replication precede autopoiesis (chapter 1)? Are the dialectical laws of historical materialism more fundamental than the biological laws of genetic replication (chapter 2)? Does nature trump nurture—or not (chapter 3)? Bioaesthetics, too, engages and thinks through these questions, but it is not defined by them. Bioaesthetics is not a discipline based on first principles or the logic of linear causality. It is neither philosophy nor science. It is the discourse of making sense.

I have revisited the intricate relation between concepts and objects not because I hope to reignite the science wars of the 1990s, but because I want to emphasize up front that metaphors and concepts *matter,* in biology as much as in the humanities. In a recent study, for example, Thierry Bardini has demonstrated the degree to which the current understanding of genetics has been shaped by the intense debate in the 1970s about the proper name to bestow on the noncoding part of the DNA, which was originally called "excess DNA" or "nonsense DNA" prior to being designated as "junk DNA." The lingering semantic ambiguity about what, exactly, scientists mean when talking about genes—whether they refer to *any* sequence of DNA or only to its *coding* sequences, including the question about whether, and how, the first interacts with the second—cannot simply be dismissed as regrettable side effects of linguistic arbitration, but constitutes a significant part of today's genetic theory at large.[57] "It appears thus quite ironic," Bardini concludes, "that the acme of today's big science, some three centuries after the modern scientific revolution, ends up producing a controversy where authority over metaphor is crucial."[58]

Before moving on, I want to provide another example to demonstrate why *bioaesthetics* must pay close attention to the production of meaning at all levels, from the molecular to the metaphorical. A key question in biology obviously concerns the relation between organic life and cognition, the topic of my next chapter. As one might expect, there is considerable tension on this point between scientific realists, on the one hand, and vitalist philosophers such as Gilles Deleuze, on the other. In some of his early writings, Deleuze refers to virtual being (or what he will later call the BwO) in its preindividual, embryonic state as a "passive self" or "larval subject" that subsists on the microbiological level, making use of philosophical concepts (self and subject) usually reserved for highly developed animals or humans exclusively.[59] His point is that life emerges from spatio-temporal dynamisms of pure affect, and that any organism, no matter how small, creates a unique space-time continuum to separate itself from its environment. Most bacteriologists, however, reject the notion that single-cell organisms are sentient beings or have a self to speak of, let alone any kind of subjectivity. If this idea is mentioned in biology textbooks at all, John Protevi points out, it is done mainly to dismiss it. In his close reading of these texts, however, Protevi reveals that scientific accounts frequently rely on phenomenological terms and first-person perspectives in trying to explain how bacteria "experience" their particular environment—even though the main point of these explanations, of course, is to deny that bacteria have experiences in the first place. This tension between the implicit and explicit meaning of scientific texts, Protevi argues, cannot be relegated to mere semantics or improper word use. Rather, "the instability of this discourse, its shifting from third-person to first-person perspective," he insists, is symptomatic of a larger problematic in scientific theory and practice, namely, that the linguistic description of scientific experiments interacts with the sense we make of them.[60]

This problem is well known in the philosophy of science. It results from the clash of two explanatory modes: the external, "linguistic mode" of describing a particular experiment, on the one hand, and the internal, "dynamic mode" that sustains the experiment itself, on the other. Although these two modes are not necessarily contradictory, they nonetheless function according to different symbolic rules, the combination of which may, and often does, result in semantic ambiguity or misunderstandings. The claim, for example, that DNA "codes for protein" is an oversimplified linguistic description of a dynamic and extremely complex biochemical process that consists of numerous steps in between:

the transcription of DNA into mRNA (messenger RNA); the subsequent excision of junk DNA; the splicing together of the remaining protein-coding DNA stretches; the binding of tRNA (transfer RNA) to mRNA; and so on.[61] As Evan Thompson points out, there is nothing wrong with using the linguistic mode as long as we remember its primary purpose, which is to abstract from the minute details of molecular processes to arrive at a general account of how DNA replication works. "Yet it is unacceptable," Thompson continues, "to say that DNA contains the information for phenotypic design, because this statement attributes an intrinsic semantic-informational status to one particular type of component and thereby divests this component of its necessary embedding in the dynamics of the autopoietic network. It is this network in its entirety that specifies the phenotypic characteristic of a cell, not one of its components. . . . In summary, the linguistic mode is emergent from the dynamical mode, and information exists only as dynamically embodied."[62] The necessity to translate dynamic modes of experimentation into linguistic modes of description inevitably introduces the kind of semantic tensions Latour, Protevi, and others have diagnosed in any number of scientific texts. This means that science needs hermeneutics no less than it needs technical apparatus and theoretical paradigms.

To support this claim, let me briefly situate today's biologism within a larger institutional and intellectual context, which turns out to be far more complicated than biologism would have us believe. As Michael Ruse, editor of *The Oxford Handbook of Philosophy of Biology* points out, there is little, if any, conceptual coherence within the various strands of the life sciences, which raises the question of what exactly binds them together into the singular discipline called "biology."[63] Typically housed in large departments, the biological sciences largely consist of semiautonomous subdisciplines that operate in relative isolation from one another, using different practical and conceptual tools to conduct their research. Their methodologies range from statistics and advanced mathematics all the way to philosophical questions about the origin and meaning of life. And despite their overall success—or rather because of it—the biological sciences are internally split among numerous areas of study (animal vs. plant sciences; molecular biology vs. macrobiology; etc.) that provide different answers to core theoretical questions (about reductionism vs. holism; individual selection vs. group-selection; etc.). Indeed, the academic history of biology—from Linnaeus's stratified taxonomy in the eighteenth century to the holistic approach of today's systems biology—demonstrates the terminological and methodological heterogeneity that has characterized the modern study of *bios.*

Hence we must not assume that the meaning of "bio" in disciplines such as "biophysics," "biochemistry," or "bioengineering" is exactly the same as that in terms like "bioart," or "bioethics," or "biophilosophy." The current discussion of "biopolitics," for example, is marred by so many diverse and conflicting views that the concept has become little more than an academic "buzzword." [64] Biology, of course, is the study of life. But how are we to conceptualize this multifarious "life" omnipresent in academic studies today? To this date, there exists no scientific agreement about what exactly life is, because there "is no generally accepted definition of life," as scientists continue to point out.[65] Erwin Schrödinger, for example, tackled the problem as early as 1944 in his seminal work *What Is Life?*[66] Yet Schrödinger's book never answered the fundamental question stated in its title. Instead, he focused on the physico-chemical laws of *inheritance*—much like Darwin never answered the crucial question about *The Origin of Species* either, but instead described their evolutionary *development* over long periods of time. Schrödinger, to be sure, did define the living cell as "an aperiodic crystal."[67] Yet he was unable to specify the scientific laws that supposedly define its essence, a task he explicitly entrusted to future generations of scientists to accomplish.[68] Schrödinger's belief in the ultimate explanatory power of scientific reductionism anticipates similar claims made later by Francis Crick, Jacques Monod, E. O. Wilson, and many other notably biologists in the twentieth century—never mind that none of them ever delivered on their promise.[69]

Schrödinger did, however, provide an extremely influential explanation for the chemical basis of biological heredity. Prior to the discovery of DNA in the early 1950s, Schrödinger already defined genes as both "law-code and executive power—or, to use another simile, they are architect's plan and builder's craft—in one."[70] This idea—that genes have a dual nature, meaning they are both (passive) information and (active) agents in one and thus solely responsible for biological reproduction—came to be known as the "central dogma" of molecular biology; it also inspired Crick and Watson's metaphorical description of DNA as the "book of life."[71] Not just speculative philosophers, but numerous scientists, too, have since called into question these and similar metaphors at the heart of what they call "the gene-myth"—that is, the popular yet scientifically dubious notion that genes function as autonomous agents solely responsible for replicating the "code" or "template" of organic life. As Michael R. Hendrickson, the director of surgical pathology at Stanford University Medical Center, points out, "Genes are undeniably an essential developmental resource, but they are only one of a number of factors that jointly determine phenotype and behavior" of living organisms.[72]

One might conclude from all this that the biological sciences have yet to come to terms with even their most basic concepts. While this is true, the problem is certainly not caused by a lack of disciplinary rigor, but emerges instead from the basic incommensurability of linguistic modes of signification, on the one hand, and the phenomenon of life, on the other. The problem, once again, is that questions like "What is human nature?" or "What is life?" appear to be meaningful when, in fact, they are not. "The question 'What is Life?' is thus a linguistic trap," Margulis and Sagan rightly point out. "To answer according to the rules of grammar, we must supply a noun, a thing. But life on Earth is more like a verb. It repairs, maintains, re-creates, and outdoes itself."[73] Life, in other words, makes sense only to the living, because its essence is irreducible to a third-person, "objective" perspective.

The impossibility to define life in scientific terms explains why today's discussion about the "biological nature" of things is as confused as the one about their "social construction" ever was—a point repeatedly raised by influential critics such as Evelyn Fox Keller, Judith Roof, Lily E. Kay, and many others.[74] The key problem, as noted earlier, is that concepts, not unlike life itself, constantly undergo "substantial mutation of sense and value" over space and time.[75] These conceptual mutations inevitably give rise to troubling metaphorical constructs—from genes to memes, from "the mental architecture of the mind" to "the universal acid of Darwinian theory." Keller emphasizes "the slippage of meaning, or polysemy, of our basic vocabulary," in particular the "morass of linguistic and conceptual vegetation" and the "tangle of meanings" that has grown around the nature/nurture debate since Darwin.[76] Although biological metaphors like "the code of life" provide "a rich source of study for students of the language or rhetoric of science," as Michael Ruse puts it, they also remind us that semantics and syntax are not just extraneous or accidental, but seminal to how we conceptualize and make sense of scientific knowledge.[77]

Bioaesthetics does not try to define "human nature" or the essence of "life." My goal is more modest and empirical in nature—namely, to develop a rudimentary semantic understanding of the ubiquitous prefix "bio" encountered in so many academic fields and disciplines today. What does "bio" mean? To begin with, the prefix "bio" signals, above all, a particular focus on the study of relations. By this, I refer not only to the institutional relations among the various disciplines that constitute the life sciences, but more importantly to their shared methodological emphasis of the structural coupling and reciprocal interrelation of organism and environment. The numerous case studies of scientific debates presented throughout the book will confirm that there simply can be no "biology"—no scientific discourse, no laboratory practice—outside this relationality.[78]

Second, the prefix "bio" also indicates a methodological commitment to move beyond the twentieth-century ideal of scientific reductionism and toward the twenty-first-century theory of emergence.[79] The study of life is essentially a process-oriented science focused on the emergence of the new: new genetic mutations, new organisms, new species, and the emergence of new life as such. Indeed, ever since the term "biology" was first introduced to designate the study of life around 1800, biologists struggled to defend the epistemological independence of their newly founded discipline: "To say that physiology is the physics of animals is to give an extremely inexact idea of it. I would as much like to say that astronomy is the physiology of the stars," the great French physiologist Xavier Bichat complained in 1800.[80] As this quote demonstrates, Bichat's protest focused on the use of scientific terminology, and he specifically objected to the unwarranted transference of (geo)physical concepts to the biological study of living organisms. His polemic suggestion to reverse this order and to reconceptualize physics in biological terms highlights not only the methodological incompatibility between these two disciplines, but also the epistemological power of conceptual language to (pre)determine the specific methods, goals, and objects of future scientific inquiry. The decisive question, Bichat recognized, is which discipline and what concepts will be allowed to define, literally, the terms of the debate and thus determine the epistemological foundation of the emergent life sciences.

Commenting on Bichat's quote some 150 years later, Georges Canguilhem went even further. He not only reaffirmed biology's independence from physics, but also wondered whether biology had not been too modest in asserting its own perspective on matter: "If one is to assert the originality of the biological, this must be in terms of the originality of one realm over the whole of experience and not over islets of experience. In the end, classical vitalism sins, paradoxically, only in its excessive modesty, in its reluctance to universalize its conception of experience."[81] Let me reiterate again that there is no reason for bioaesthetics to join this debate. For regardless of how we construe the disciplinary hierarchy between physics and biology, or between the humanities and the sciences, what matters epistemologically is that each is irreducible to the other. Each academic discipline constitutes a distinct system of conceptual tools and discursive practices. For that, precisely, is their purpose: each discipline, by its very nature, is charged with presenting a substantially different view of how things are. This is why disciplinary knowledge is so densely focused and well defined in its center yet becomes increasingly fuzzy toward the edges of the field. If bioaesthetics—or biology, for that matter—were ever to embrace scientific reductionism tout court, it would simply cease to exist.[82]

It follows, finally, that both biology and bioaesthetics are essentially historical disciplines: "Nothing, I repeat," says Steven Rose, "makes sense in biology except in the context of history—and within history I include evolution, development, social, cultural and technological history."[83] Let me emphasize again that these three methodological principles—first, the structural coupling of organism and environment, of objects and concepts; second, the focus on emergent properties and the limited usefulness of scientific reductionism; and third, the orientation toward historical processes and evolutionary change—are by no means meant to define the discipline of biology. They are merely meant to help clarify the basic meaning of the prefix "bio" shared by numerous fields and disciplines across the disciplines today. The first three chapters in particular will develop each of these three principles in more detail.

A similar semantic clarification, I believe, is called for with regard to "aesthetics," the second part of my neologism. In terms of its key concepts the discipline of aesthetics, like that of biology, has become a victim of its own success. As the editors of the seven-volume encyclopedia entitled *Ästhetische Grundbegriffe* point out, the term "aesthetics" has undergone a radical "delimitation of its concept" over the last few decades, and is currently applied in a wide variety of contexts that threaten to vaporize its meaning. The "particularization of aesthetics into 'genitive aesthetics,'" they argue, "has progressed to such a degree that there seems hardly a single aspect that could not be aestheticized."[84] Their overriding concern is that this proliferation of genitive aesthetics threatens to diminish rather than increase its relevance for the study of contemporary culture. Mindful of this problem, I propose to understand "aesthetics" in the context of this study as a discourse of interdisciplinary mediation organized around the three principles outlined above (i.e., structural coupling, limited reductionism, and historical change). Similar to biology, aesthetics is a meta-discipline: it enables hermeneutic "diagnoses of the connections between technical developments and our mode of sensing, experiencing, perceiving" the world around us.[85] Like any other (medical, scientific, political, historical) diagnoses, aesthetic diagnoses are more than just objective descriptions of how things are, because they actively intervene and shape their objects of study. Bioaesthetics does not just *describe* the "operative coupling and interconnectivity" within the discursive network of the life sciences; it also *functions* within these discourses and hopefully affects the knowledge they produce.[86]

Taken together, the terms "bio" and "aesthetics" signify an interdisciplinary approach to the study of culture that moves beyond the speculative theory of art that has dominated the humanities since the early nineteenth century, without, however, succumbing to the broad universalist claims that characterize today's biologism and neo-Darwinian theory. Moving beyond sociobiology and evolutionary psychology, bioaesthetics is informed instead by more recent—and more productive—approaches toward human cognition and culture developed by Francisco Varela, Terrence Deacon, Evan Thompson, John Protevi, George Lakoff and Mark Johnson, and Margulis and Sagan, among others. Hence the subtitle of my book—*Making Sense of Life in Science and the Arts*—cites a key term from "enactivism" and "embodied mind" theory to signal a bottom-up approach in evolutionary biology and cognitive science that explains human rationality in terms of basic motor function and physical movement in space.[87] Simply put, making sense is what organisms do throughout life to stay alive. This model of embodied cognition makes sense to bioaesthetics, because it reconnects body and mind, emphasizes the neuronal plasticity of the human brain, and spells out the evolutionary importance of sociocultural learning for human beings.

Bioaesthetics is conceived as a counter-paradigm to sociobiology and an alternative to E. O. Wilson's notion of consilience. This might seem like a lofty target, but I firmly believe bioaesthetics will live up to the task. Its overarching goal is to study concepts and objects in different milieus under varying historical constellations. On the conceptual side, these milieus include scientific theories and practices from multiple disciplines (e.g., evolutionary biology and neuroscience, psychology, philosophy, science and media studies). On the material side, these milieus include art objects and cultural phenomena of all kinds, from paintings and poetry to the modern mind and technological apparatus. The proper *Umwelt* of bioaesthetics is not one, but many. The milieus it studies are manifold and complex, and they range from intracellular, molecular environments to the global economy of international art markets and digital culture. No single theoretical framework is sufficient to foster this kind of analysis, and only a multiplicity of local perspectives can help elucidate its conceptual and objective terrain.

Hence the only way to foster genuine interdisciplinary dialogue is to compare and contrast leading scholars and research paradigms usually studied in isolation. Most cultural anthropologists and evolutionary psychologists, I have come to conclude after years of study, are entirely unfamiliar with current media theory and Bernhard Stiegler's influential theory of "technogenesis," and vice versa—despite the methodological and conceptual ties that connect the two disciplines (see chapter 3). Similarly, the institutional disconnect between analytical and continental philosophy has led cognitive and cultural theorists in the social sciences to disregard the affective turn in the humanities along with the vitalist philosophy of Gilles Deleuze—and vice versa (see the Coda). Bioaesthetics, for sure, cannot possibly mend these gaps either, nor does it try to. But it can, and it does, measure the distance and establish basic communication within and between some of these gaps in an exemplary fashion. Which is to say that bioaesthetics falls far short of consilience, yet is all the more productive because of it.

Structure and Chapters

Including the Introduction and the Coda, this book consists of seven chapters. Each one begins with a particular event and introduces one or several important debates in the history of bioaesthetics, starting with the eighteenth-century debate about preformationism and epigenesis and working its way forward to the ongoing controversy about sociobiology and neo-Darwinian theory. In the course of these debates we will encounter the prehistory of biologism, and we will study how scientific concepts gain meaning and (fail to) make sense. As mentioned above, the first three chapters focus in particular on the three main principles that inform bioaesthetics. Reading Kant in the context of autopoietic theory, for example, I argue that his transcendentalism amounts to what biologists call the "structural coupling" of organism and environment. This means that living systems are inextricably tied to the specific environment in which they operate. But these systems do not find their environment waiting for them ready at hand; instead, they create their unique environment much like the environment creates living systems. There is no linear temporal order to this process, Kant realized. Life defies human logic.

Kant's philosophy is neither constructivist nor realist, but correlational: it thinks through the simultaneous emergence of reason and reality, concept and object. The ensuing paradox explains the appearance of so many temporal contradictions in Kant's *Critiques*, in particular those related to aesthetic judgments (e.g., What comes first: The feeling or the judgment? The categories or the objects they categorize?). Overall, my reading of Kant supports Rachel Zuckert's claim that the *Third Critique* envisions a new kind of human subjectivity distinct from both the transcendental subject and empirical subjects—an indeterminate subject "freely open to, and holding itself to, finding contingent, conceptually indeterminate unity in its representations."[88] This subject, I argue, is bioaesthetic. It emerges through the suspension of time that characterizes genuine aesthetic judgments. Facing beauty, Kant's bioaesthetic subject literally enacts a sense of its own being to grasp what it means to be human.

The second chapter centers on the relation of emergentism and reductionism in modern science. I situate my discussion in the broad historical context of the relation between Marx's philosophy of human history, on the one hand, and Darwin's theory of natural evolution, on the other. I focus in particular on the intriguing—though today mostly forgotten—debate between the structuralist Marxist Louis Althusser and the molecular biologist Jacques Monod in 1968 in Paris. We shall see that Althusser criticized Monod (and modern science in general) for his hidden philosophical idealism and his ignorance about the "scientific laws" of historical materialism, while Monod criticized Althusser (and Marxist philosophy in general) for his economic determinism and the anthropological illusion that man is the "necessary end-product of a cosmic process." My goal in revisiting this debate is not to settle the score, but to argue that argumentative gridlocks of this kind emerge due to the logical incompatibility of a set of first principles and "natural laws" rooted in different epistemological systems, each of which seeks to dictate its own terms to others and establish its version of the universal "law" that allegedly governs the exchange between them.

In order to demonstrate this point further, my third chapter revisits the famous sociobiology debate that started in the 1970s and continued to resonate throughout the 1980s and 1990s. The reason why E. O. Wilson's original thesis of sociobiology, much like his later theory of consilience, caused such a stir in the academy was that he promoted an allegedly neutral space of mutual respect and collaboration between the two cultures, yet proceeded to define this space exclusively in molecular biological as opposed to sociohistorical terms. Although Wilson and other neo-Darwinians (e.g., Richard Dawkins, Daniel Dennett, Steven Pinker, John Tooby, Leah Cosmides, Jerome Barkow) emphasize the relative autonomy of human cultural development, they nonetheless insist that the fundamental laws of natural selection ultimately determine these developments. Darwin's theory, as Dennett famously argued, rules across "all the achievements of human culture—language, art, religion, ethics, science itself," meaning that "life and all its glories are thus united under a single perspective."[89] By contrast, I argue that Wilson's version of consilience managed to rearticulate ancient philosophical dilemmas in such a way as to render genetic–molecular explanations inherently more plausible than any other. Throughout human history, of course, it has always been the privilege of the stronger party to choose the allegedly "neutral" language in which a peace treaty is to be couched. Its final outcome will thus have been predetermined from the very beginning, because the choice of language is always already a dictation of terms. Trying to keep the peace, the weaker party has little choice but to accept these terms both literally and metaphorically. This, precisely, is what biologistic theories of art and culture have done. If left unchecked, their influence on research in the humanities will be disastrous.

The fourth and fifth chapters of the book are meant to showcase the most pertinent examples of this disaster in current biologism. My survey, I hope, will also help clarify the distinction between biologism and bioaesthetics in more detail. To this end, I introduce and critique the numerous academic fields and disciplines (e.g., "literary Darwinism," "biopoetics," "neuroaesthetics," "cognitive cultural studies") that have emerged under the current hegemony of the life sciences. I have categorized them into two major strands. I call the first strand "evolutionary aesthetics," because it provides a broad Darwinian perspective on the evolutionary history and adaptational function of human culture in general and literature in particular. The second strand is "neuroaesthetics," which zooms in on the intersection between neuronal activity in the human brain and the emergence of aesthetic experience, focusing mostly, though not exclusively, on the visual arts. Needless to say, this binary classification is heuristic and far from absolute, meaning that other taxonomies and distinctions are certainly possible and might prove more productive in some regards, though less so in others. Taken together, the two chapters sketch a rudimentary map of the distinct—at times even mutually exclusive—conceptual frameworks and interdisciplinary relations that inform these various approaches. We shall find that most strands of evolutionary aesthetics and neuroaesthetics are either implicitly or explicitly based on Wilson's notion of sociobiology and neo-Darwinian theory. But there are also other approaches that either ignore or explicitly reject Wilson's orthodox view of cultural evolution, and others still that present a sociohistorical critique of the radical scientific reductionism that informs it.

Despite their differences, many of these approaches end up complementing one another, for the simple reason that they do not always share the same object of study. That is to say they conceptualize their object differently even if they claim to have the same referent in mind. As a general rule, one can say that the more a particular theory or discipline universalizes its results and disavows the historical constructedness of its object of study, the less innovative and productive that theory will become over time, and the more predictable and uninspiring its findings will be. Unlike biologism, bioaesthetics welcomes the methodological and conceptual diversity across the disciplines as well as the productive tension between empirical and speculative theories of culture. It recognizes that the study of art and aesthetic experience requires a broad range of diagnostic tools and analytical perspectives irreducible to one set of scientific principles or theoretical paradigms. The goal of bioaesthetics is not to police the migration of concepts across disciplinary borders, but rather to study the discursive and material effects of these migrations in order to decide whether and how they make sense.

In the Coda, I situate bioaesthetics in the context of the affective turn and Deleuze's vitalist aesthetics of becoming. The most productive reading of Deleuze, I argue, is to read him in the context of recent cognitive theories about the embodied mind and enacted being. That context includes not only the history of Darwinian evolutionary theory, but also that of Kantian transcendentalism. Ultimately, what distinguishes Deleuze's aesthetics from my notion of bioaesthetics is the Kantian emphasis on human judgment and my belief that sense making is an active rather than passive mode of existence. For Deleuze, as for Derrida, Badiou, and many other French philosophers, sense just happens when the mind idles and human habits relax their grip on our senses. For me, sense emerges not least because we *make* it happen. The two views, however, merely differ in degree, not in kind. They derive from different perspectives on what is essentially the same phenomenon of human creativity and aesthetic experience. Hence my hope that the arguments presented in this book will amount to a Benjaminian constellation of ideas or even a Deleuzian rhizome of one sort or another—except that bioaesthetics, unlike Deleuze, welcomes the organism into the fold and does not consider it the enemy. And while some Deleuzian critics today advocate a "noncritical form of reading" entirely severed from human beings, bioaesthetics emphasizes the importance of a "critical" posthumanism that must continue to inform the more "affirmative" stance favored by Deleuze.[90]

Finally, I want to clarify that this book does not provide any close readings of particular artworks or cultural phenomena. This will have to be done in a separate study, one that shifts focus from the concept-centered approaches discussed here to an object-centered analysis of particular artifacts situated within particular sociohistorical contexts. Unlike traditional aesthetics, bioaesthetics provides no guidelines as to who or what shall become the object of aesthetic investigation. But it does insist that once we have zoomed in and singled out a particular object of study, we remain aware of the analytical perspective and specific approach we employ to get the results we need. Bioaesthetics knows that whatever we see in the world is, in part, a reflection of the light we shine on it. No subject, and no object, can be conceived outside this paradox.

So whether scholars look up close at specific concepts in scientific texts or widen their gaze to survey life on earth from an anthropological perspective, it is important to remember that none of these objects or sites need be misconstrued as paradigmatic for bioaesthetics at large. Paradigms are always relative, not absolute. Rather than seeking the "universal laws" of art, bioaesthetics demonstrates that aesthetic objects always emerge as historically situated and media-specific sites of conceptual and material conflicts between multiple forces. These forces and their interactions create a local environment irreducible to the singular logic of evolutionary theory, or Marxist philosophy, or Heideggerian phenomenology, or any other totalizing epistemological framework in modern history. Instead of replacing one paradigm with another, the task at hand is to link these paradigms or make them collide to see what happens. The goal is to expose the insufficiency of one paradigm in light of the other, and thus to come to terms with paradox and the conceptual tensions always already at work in bioaesthetics.

1 HUMAN NATURE AFTER KANT

> We must conceive discourse as a violence that we do to things, or, at all events, as a practice we impose upon them.
>
> **MICHEL FOUCAULT**

Kant and Biology

There are good reasons to begin our discussion of bioaesthetics with Kant, who published at a time when academic disciplines were still being established and the "two cultures" did not yet exist.[1] Aesthetics, indeed, was a new academic discipline defined by the German philosopher Gottlieb Alexander Baumgarten in 1750 as the "science of sensory knowledge." Aesthetics, according to Baumgarten, was meant to clarify the relation between the sensory data based on human experience, on the one hand, and the scientific knowledge that could be derived from that data via the use of reason, on the other. In his *First Critique* of 1781, Kant, too, defined aesthetics as a "science," which he called "transcendental aesthetics."[2] This science provides a non-empirical analysis of the pure forms of sensibility—space and time—that determine how objects must appear to the human mind. It is important to note that transcendental aesthetics, in Kant's sense, has nothing to do with art or our experience of actual sensations. It merely outlines the a priori conditions that enable us to have sensations in the first place.

In a footnote, however, Kant took care to distinguish this transcendental (scientific) sense of aesthetics from another, merely empirical sense in which aesthetics denotes the "critique of taste."[3] Popular among British philosophers at the time, this second, empirical sense of aesthetics clearly dominates Kant's later *Third Critique* of 1789, even though it is not always easy to keep the two (transcendental and empirical aesthetics) apart. In his *Third Critique,* Kant provided a deduction of aesthetic judgments, offered a definition of artistic genius, and set up a basic topology of the arts. Still, even the *Third Critique* did not have much to say about the history of art, nor did Kant present any detailed analyses of individual art objects. His focus throughout remained on aesthetic experience and judgments, and his main examples were drawn from the realm of nature as opposed to the cultural realm of manmade artifacts.

The point I want to make is that aesthetics originally was and remains today a reception-oriented as opposed to a production-oriented discourse. It is concerned with how, and why, we perceive certain objects as pleasant or beautiful or sublime. Yet aesthetics cares less about how, and why, such objects come into being in the first place or what their material qualities and formal structures are—questions that are essential to the history and philosophy of art. Whereas Kant took care to distinguish the two senses of aesthetics (transcendental and empirical), neither of which can simply be equated with the history or philosophy of art, the later romantic tradition and nineteenth-century German idealism all but ignored these distinctions and essentially collapsed art and aesthetics into one.[4] For my account of bioaesthetics, however, it is paramount to keep them apart. Art and aesthetics are related disciplines for sure, but they are not identical by any means, because aesthetic discourse pertains not only to art, but to nature as well, and it tends to focus on the subjective (mental) rather than the objective (material) dimension of phenomenological experience.

As I will demonstrate in later chapters of this book, the failure to distinguish between art and aesthetics causes serious methodological problems for today's biologist approaches to culture. In trying to define the "nature of art" exclusively in scientific terms, biologism reduces art to knowledge and aesthetic judgments to cognitive judgments. Aesthetic experience, according to this model, is little else but "cognitive play," a welcome opportunity to practice and sharpen our ability to reason as a means to increase our chances for survival. For Kant, however, and for bioaesthetics in general, such cognitive play is *necessary but insufficient* to account for genuine aesthetic experience. The latter exceeds (or undermines) cognition to communicate a subjective feeling or *Gemütszustand* (mental state) common to all members of the human species. It is the peculiar nature of this subjective yet universally shared feeling that distinguishes genuine aesthetic judgments from both cognitive judgments, on the one hand, and mere judgments of taste, on the other. Strictly speaking, aesthetics, for Kant, is neither a science nor a philosophy, but a sense—a sense of being human, first of all, but also a sense of our inherent ability to *make sense* in and through aesthetic judgments.

As these comments indicate, Kant's philosophy was deeply informed by the empirical study of nature. His *Third Critique* in particular engaged many of the key questions raised by the bourgeoning life sciences, such as the taxonomy of the species and the emergence of life itself. Yet as early as 1763, in an effort to prove the existence of God with the help of natural science, Kant had already acknowledged the need to explain how plants and animals grow. Organic growth, he argued, was either "a divine act, or an ability *[Tauglichkeit]* is to be granted to the first divine arrangement of plants and animals to generate their likeness in temporal succession *[in der Folge]* according to a natural law, not purely by development *[entwickeln]*, but truly by procreation *[erzeugen]*."[5] This early conceptual distinction between the genuine "procreation" versus the mere "development" of organic life and human knowledge continued to play a major role throughout Kant's later critical philosophy. At the end of his *First Critique*, for example, Kant compares the "architectonic structure of pure reason" to the natural growth of an animal body *(ein tierischer Körper)*. This analogy serves to emphasize the dynamic–organic rather than static–mechanical nature of our ability to reason. This ability, Kant suggests, grows organically from within and follows its own formative drive analogous to how living organisms develop. Whereas the "technical unity" of empirical–scientific knowledge consists of a mere aggregate of disjointed pieces, the "inner, organized unity" of reason—and, by extension, the unified structure of Kant's own transcendental philosophy—comprises coevolving parts that are both means and end, both cause and effect of the larger whole. Unlike machines, yet akin to living organisms, human reason is a "self-organizing" structure able to assemble and maintain itself: "Thus the whole is organized *(articulatio)*, not accumulated *(coacervatio)*. It may grow internally *(per intus susceptionem)*, but not externally *(per apposotionem)*, like an animal body, the growth of which does not add any new member but, without changing their proportion, renders each member stronger and more efficient for its ends."[6] To clarify this point, Kant elsewhere in *Critique of Pure Reason* refers to the "augmentation of concepts from themselves, and, so to say, the self-birth of our understanding (and of our reason), without impregnation by experience."[7] The power of human understanding/reason thus derives from—yet, at the same time, also gives rise to—fundamental laws (rules or principles) that emerge solely during the self-conscious act of reasoning itself. Kant emphasizes that these basic laws are neither innate in the mind nor acquired through experience, but spring forth spontaneously and necessarily whenever "I think" (an activity Kant calls "transcendental apperception"):

> By analytic of concepts I do not mean their analysis . . . , but a hitherto seldom attempted analysis of the faculty of the understanding itself, in order to investigate the possibility of *a priori* concepts by looking for them nowhere but in the understanding itself, as their birthplace, and by analyzing the pure use of the understanding in general. This is the proper task of a transcendental philosophy. . . . We shall therefore follow up the pure concepts to the point where they first sprout and show their dispositions in the human understanding, in which they lie ready, till at last, occasioned by experience, they become developed, and are exhibited by the same understanding in their purity, freed from all adhering empirical conditions.[8]

"Like an animal body," "birthplace," "dispositions"—these and other biological terms and metaphors clearly demonstrate the importance of the bourgeoning life sciences for Kant's critique. In the second edition of *CPR* from 1787, Kant drives this analogy between (his critique of) reason and (the genesis of) life even further. In a famous passage, Kant now refers to "the epigenesis of pure reason," using a term that was as central to eighteenth-century debates in embryology as it is in genetics and evolutionary biology today.[9]

Kant, of course, knew nothing about Darwin and modern evolutionary theory, let alone molecular genetics or complex systems. Even the term "biology" was not yet used to designate the scientific study of living nature prior to the nineteenth century.[10] And Kant himself repeatedly emphasized that studies about the genesis and development of life, important as they were, could not possibly amount to the status of "proper science"—a title he reserved for mathematics and Newtonian physics alone.[11] "It would be absurd for humans . . . to hope that there may yet arise a Newton who could make comprehensible even the generation of a blade of grass according to natural laws that no intention has ordered; rather, we must absolutely deny this insight to human beings," Kant famously claimed in his *Critique of Judgment*.[12] Kant adamantly rejected the idea of "hylozoism"—that is, the notion "that life should have arisen from the nature of the lifeless, and that matter should have been able to assemble itself into the form of a self-preserving purposiveness by itself."[13] Instead,

Kant insisted that life—what he called "organized entities" or "natural purposes" (*organisierte Wesen* or *Naturzwecke*)—must be considered a primary substance irreducible to inanimate matter.

In the course of this chapter, I will explain in more detail why "organized entities," according to Kant, must be conceived as purposive (teleological) beings. It has to do with the limited faculty of our understanding, which only recognizes mechanical or teleological causes to explain how things come about, including life. And since the self-organizing complexity of living organisms cannot be explained via simple mechanical principles, there must be a teleological principle at work—this, at least, is what reason forces us to conclude. This conclusion, however, is hypothetical and impossible to verify empirically. In fact, we shall see that Kant himself ardently believed it was wrong. Still, the only way to move forward, he argued, was to recognize the logical necessity to proceed "as if" life were teleological, while, at the same time, to remain agnostic about whether this is actually the case. Kant considered life categorically distinct from matter. He also thought that human beings were differed in kind from other animals and that the human soul is irreducible to our body. This strict bifurcation of animal and human nature, along with Kant's claim about the purposeful, teleological nature of "organized entities," clearly puts him at odds with today's (Darwinian-based) biology, according to which humans *are* animals, yet life evolves *without* purpose or telos of any kind.

Given these discrepancies and the historical distance that separates Kant's time from our own, how are we to understand his claim about the "epigenesis of pure reason"? What does this term mean in the historical context of the bourgeoning life sciences at the end of the eighteenth century, and how is this debate relevant for our concept of bioaesthetics? And why, despite Kant's apparent lack of evolutionary theory, do historians and philosophers of science today insist that he remains "a genuine forerunner of investigations into 'epigenetics' and 'emergent properties' of genes" in the twenty-first century?[14]

To address these questions, I will first provide a brief overview of the two major theories regarding the origin and development of life during Kant's time, namely, preformationism and epigenesis—although one must add that this binary terminology is somewhat misleading, given the many nuanced variations between these two poles. Still, it seems fair to say that preformationism sought to provide a purely mechanical explanation of life, whereas epigenesis relied on various kinds of vitalism to make the case for the irreducibility of life to mechanical causes. Although Kant ultimately adopted the term "epigenesis" to describe his own position in 1787, he actually tried to develop a third alternative—a middle path—situated between mechanism and vitalism. We shall see that Kant's theory closely resembles the concept of autopoiesis developed by Humberto Maturana and Francisco Varela during the 1970s and 1980s, which explains how living systems constitute themselves by means of organizational recursivity: "The most striking feature of an autopoietic system," Maturana and Varela point out some two hundred years after Kant, "is that it pulls itself up by its own bootstraps and becomes distinct from its own environment through its own dynamics in such a way that both things are inseparable."[15] Their key point is that autopoietic processes determine not only the physiological structure, but also the cognitive abilities of living systems, because knowledge is always and necessarily recursively organized: "Knowing how we know does not consist of a linear explanation that begins with a solid starting point and develops to completion as everything becomes explained. . . . The business of living keeps no records concerning origins. All we can do is generate explanations, through language, that reveal the mechanism of bringing forth a world. . . . As we know how we know, we bring forth ourselves."[16] Living organisms, in other words, are both autopoietic and cognitive systems.[17]

Kant knew that, too. His philosophy, I argue, anticipates Maturana and Varela's central claim that living organisms are self-organizing systems, and that their cognition remains inextricably tied to the particular environment they enact by virtue of being alive. This insight, I believe, is why Kant referenced the self-organizing complexity of living bodies as a means to elucidate the "inner, organized unity" of human knowledge and transcendental philosophy. In the last section of this chapter, I argue that biological problems and concepts inform not only Kant's epistemological critique, but his aesthetic theory as well. The "bio" in Kant's aesthetics is not only evident in his discussion of "epigenesis" in the *Third Critique*, but also informs his understanding of the *sensus communis* as an essential component of human nature informed by aesthetic experience.

Preformation and Epigenesis before 1800

Until the mid-eighteenth century, the dominant theory to explain the existence of life was preformationism.[18] The most extreme or "strong" version of preformationism—often referred to as "preexistence" or "encasement" theory—was formulated by Malebranche in 1674, who argued that all living organisms had been conceived by God at the time of creation and have continued to exist since then, nestled one within the other like Russian dolls waiting to be released.[19] By the end of the seventeenth century, these preexisting organisms were commonly conceived as tiny homunculi or animalcules—that is, miniature yet fully formed versions of individual life forms—located in either the female ova or the male spermatozoa. Either way, the core assumption shared by all theories of strong preformationism was that each and every individual life form had originally been created by God as a whole and morphologically complete organism at the beginning of time. Although these preformed individuals required the act of conception to be released, their maturation process was purely mechanical and restricted to size, not form. Conception, in other words, did not add anything new to the generative process of preformed individuals. It merely triggered the release of what had always already existed in toto. According to strong preformationism, life *unfolded,* rather than genuinely *developed,* over time—although critical terms like these were often used interchangeably, a problem due, at least in part, to mistranslations from one language to another.[20]

A second set of theories, unified mainly by their shared rejection of strong preformationism, was called epigenesis. In his voluminous and highly influential study *Histoire naturelle, génerale et particulière* (1749–67), Georges Buffon introduced the flow of time into the static and ahistorical taxonomy of living forms developed by Carl Linnaeus, the Swedish botanist, in his *Systema Naturae* from 1735. Linnaeus's system was gospel to the preformationists, who considered it empirical evidence in support of their theory. According to Buffon, however, the key to understanding life was to acknowledge the natural agency of "living matter" to change over time. Whereas the preformationists assumed that each and every form of life classified in Linnaeus's table had preexisted since the beginning of time, Buffon argued that many of these forms were not original, but had actually developed on their own since then. This historical development of living forms, Buffon claimed, was due to the "active principles" or "organic particles" *(molécules organiques)* God had created together with the original form—or what Buffon called the "internal mould" *(moule interièure)*—of the first individuals of each species:

> God, when he created the first individuals of each species of animal and vegetable, not only bestowed form on the dust of the earth, but gave it

> animation, by infusing into these individuals a greater or smaller quantity of active principles, of living organic particles, which are indestructible and common to every organized being. These particles pass from body to body, and are equally causes of life, of the continuation of the species, of growth, and of nutrition. After the dissolution of the body, after it is reduced to ashes, these organic particles, upon which death has no influence, survive, circulate through the universe, pass into other beings, and produce life and nourishment.[21]

With the death of each generation, the species develop further and further away from their original form, because they must rely on the organic particles to re-create the unique specificity of their internal mould. According to Buffon, this successive degeneration was caused by environmental changes such as new foods and different climates that adversely affected the living particles and, over time, degraded their ability to replicate faithfully the species' internal mould in later generations—hence the multiplicity of seemingly unrelated forms in living nature. Buffon speculated that this degenerative process might be reversible, at least in principle, provided one were able to determine and re-create the original environmental conditions encountered by the first phenotype of each species at the beginning of time.[22]

The divine origin and fixed internal form of the species, along with the reversible nature of their historical degeneration, clearly distinguishes Buffon's theory from the genuine evolution of the species discovered by Darwin a century later. Since Darwin, it literally makes no sense to speak of the historical *degeneration* of species, because there is no preset norm—no original form to maintain—in the ongoing evolution of life. In the eighteenth century, however, Buffon's theory was extremely influential. Apart from challenging Linnaeus's static taxonomy, its success was also due to the more satisfactory account Buffon provided for the role of conception. From early on, preformationists had had difficulties explaining the fact that children clearly exhibited traits inherited from not just one, but both parents. For according to their theory, each individual offspring was already fully formed in either the female ova or the male sperm, leaving nothing to contribute for the other parent. Buffon, on the other hand, argued that sexual reproduction consisted of putting together the organic molecules from both parents—not, to be sure, by merging them into an altogether new form, but simply by putting two preformed pieces together side by side to complete the whole.

The central remaining question in Buffon's theory, however, concerned the specific nature of the force necessary to animate the particles. What, exactly, was it and from whence did it come? Strongly influenced by Newton, Buffon conceived of this "penetrating force" in analogy to that of gravity: both were immaterial forces that could only be known and studied indirectly, in terms of their material effects.[23] Like gravity, the penetrating force was essentially a non-teleological, purely mechanical force that had nothing to do with Aristotelian entelechy, for example. It was not a "vital" force of any kind, but a blind mechanism that organized living matter. "Life," Buffon claimed, "far from being a metaphysical degree of being, is a physical property of matter."[24] Yet unlike the force of gravity, which applies to all forms of matter, Buffon's penetrating force operated exclusively within "living things" as opposed to nonorganic, unorganized bodies. Though mechanical, the penetrating force was unique to living matter and inseparable from it. Without this force, Buffon surmised, the organic particles would no longer be able to replicate life.

Buffon's critics, particularly in Germany, were quick to challenge the dubious nature of this mechanical force allegedly inherent within matter itself.[25] In his introduction to the German translation of Buffon's work in 1751, the renowned botanist Albrecht von Haller, professor in Göttingen, went straight to the heart of the mater: "If matter has forces that allow it to build things, it does not possess them blindly," Haller insisted. Otherwise, there was no way to explain the heterogeneity of organic life forms nor their high degree of internal complexity, the organization of which far exceeded the mechanical laws of nature. The complex diversity of life simply exceeds "the constraints of blindly working matter."[26] A similar critique of Buffon's penetrating force was still advanced by Caspar Friedrich Wolff toward the end of the eighteenth century: "There would be no reason why this force would work now in one way, now in another," Wolff wrote in 1789, "nor why it would produce at one time one part and at another time another."[27]

Although Wolff's and Haller's critiques of Buffon were essentially the same—namely, that a pure mechanical force cannot possibly explain how organized life emerges from matter—they suggested different theories to fix the problem. Wolff went on to reject all theories of preformationism in favor of epigenesis and what he called life's "essential force" (*vis essentialis* or *wesentliche Kraft*), whereas Haller moved into the exact opposite direction. "Nature," Haller concluded, "only copies moulds already created."[28] The only way to account for the constancy of species development, Haller insisted, was to return to the preformationist theory of ovism—albeit with the caveat that God had not preformed every single individual in the female ovum, as the strong version of preexistence falsely assumed. Instead, God had merely created generic models or "preformed germs" *(vorhergebildete Keime)* for each species that would unfold over time:

> It appears very probable to me that the essential parts of the fetus exist formed at all times; not, it is true, in the way that they appear in the adult animal: they are arranged in such a fashion that certain prepared causes, hastening the growth of some of these parts, impeding that of others, changing positions, rendering visible organs which were formerly aphanous, giving consistency to the fluidity and to the mucosity, form in the end an animal which is very different from the embryo, and yet in which there is no part that did not exist essentially in the embryo. It is thus that I explain development.[29]

Given Haller's emphasis of preformed germs rather than preformed individuals, his theory came to be known as "generic preformationism" in order to distinguish it from the strong version of "individual preformationism" described above. This distinction, obviously, is important, yet the specific terms used to designate it have confused rather than clarified the issue, in my view. Buffon, for example, is usually called a "mechanical epigenesist," because he explicitly rejected preformationism and emphasized instead the historical change of organic matter due to mechanical forces and the environment. This emphasis of mechanicism, to be sure, distinguishes Buffon's theory of epigenesis from other, more vitalist versions like those of Caspar Friedrich Wolff, Johann Friedrich Blumenbach, and Johann Gottfried Herder. At the same time, however, Buffon's "mechanical epigenesis" might just as well be classified as yet another kind of "generic preformationism." After all, both the internal mould and the organic particles of each species, according to Buffon, had been preformed and created by God at the beginning of time. Although living forms degenerate over time, they do not really "develop" or "adapt" in the Darwinian sense of acquiring new and permanent traits. Strictly speaking, there is no evolution or "genesis" of any kind in Buffon's theory. Whatever change or degradation organic particles endure throughout history always remains reversible in principle, because all existing forms of life can be traced back to their original mould bestowed by God. Such a theory, I think, can aptly be described as a kind of "generic preformationism."

My overall point is that labels like "mechanical epigenesis" or "generic preformationism" can do more harm than good, because they provide insufficient, and frequently misleading, shorthand descriptions of complex theories that defy simple means of classification. The primary function of these labels, I would argue, is polemical, because they allow scholars to associate themselves with this or that particular academic field so as to rally support against their academic opponents from other camps. And more often than not, the closer a rival theory seems to mirror one's own, the harsher the rhetoric becomes. That is because scientific debates are not just about data. They are about the meaningful *interpretation* of data within a larger conceptual framework. That framework, however, is subject to historical developments and sociopolitical forces that often have little to do with "proper" science. National pride and personal ambition, as well as the need to secure the necessary funding for ongoing research, have shaped the modern history of academic debates no less than the scientific experiments that generated them in the first place.

Take Haller, for example. He certainly went to great length to distinguish his model of "generic preformationism" from Buffon's theory of "mechanical epigenesis." Buffon's theory, Haller argued, was forced to rely on some inexplicable "vitalist" force in order to explain the organic development of life—never mind that Buffon himself repeatedly and passionately denied that charge. Yet Haller's own theory accomplished little more than restating the original problem in different terms. All he did, essentially, was to shift the focus once again back on life's preexisting *forms* as opposed to the (vital or mechanical) *force* needed to animate them. This analytical shift, however, did not solve the conundrum of effective causation in living matter. For what, exactly, are "the essential parts of the fetus" that, according to Haller, "exist formed at all times"? And what is the precise nature of those "prepared causes" that, in the end, form "an animal which is very different from the embryo"? Haller claimed, for example, that the heart was the essential organ of the fetus charged with assembling all others parts and directing the growth of the entire organism. Yet this claim was demonstrably false, as his rival Wolff and others soon pointed out. Their empirical studies proved that the heart grows at the same rate as the rest of the organism, hence it could not possibly be in control of organic development overall.

It was this ongoing, intense debate about the origin and development of living organisms that galvanized Kant's long-standing interest in natural history and prompted him to clarify his own position on the matter. Although he was critical of Haller's preformationism, Kant nonetheless adopted the term "generic preformation" for his own purposes and equated it with (what he understood as) "epigenesis." Obviously, this conflation of terms further increased the already considerable confusion about their respective meanings both during Kant's time and for us today. Yet the fact that Kant used both terms interchangeably serves to indicate what he sought to achieve in his own work—namely, to reconcile and find some middle ground between epigenesist and preformationist theories of life. Kant's first step was to augment Haller's theory of preformed germs. In his essay "Of the Different Human Races" from 1775, Kant distinguishes between *Keime* (germs) and *Anlagen* (predispositions). Both of these, he argued, must work together to determine, first, the qualitative growth of individual body parts, and second, their successful assemblage into a well-structured organism able to survive and reproduce. According to Kant, *Keime* are the preexisting seeds that determine the distinct parts and general physical properties of each species, whereas the *Anlagen* are triggered by the particular environmental conditions encountered by the phenotype in order to determine the optimal size and structural relationship of these parts to one another.[30]

Kant's modification of Haller's theory resulted in a "combination of preformationism with environmentalism," as Phillip Sloan points out.[31] Kant frequently used this model to support his arguments in both the *First* and *Second Critique,* for example in the famous passage from *CPR* cited above, in which he suggests that we must "follow up the pure concepts to the point where they first sprout and show their dispositions in the human understanding, in which they lie ready."[32] Moreover, Kant's "environmentalism" was clearly influenced by Buffon, who had been the first naturalist to consider environmental changes responsible for the historical development of living forms. In Buffon's theory, however, this process was merely degenerative, whereas in Kant's theory, it was truly epigenetic. The natural predispositions within the organism, Kant argued, enable each phenotype to adapt—within preestablished limits—to the "future circumstances" and possible "changes with regard to climate or soil."[33]

In his anthropological studies, for example, Kant discerned four distinct races within the human species, the different features and skin tone of which, he argued, were the historical result of generic adaptations necessary for each race to survive within its unique environment. What explains the existence of racial differences, in his words, was the "foresight" (*Vorsorge,* 1st ed.) or "care" (*Fürsorge,* 2nd ed.) of nature trying to protect its own creations via in-built adaptations triggered by different environments.[34] Despite their differences, however, Kant nonetheless insisted that all human beings, regardless of race, must have descended from one original *Stammgattung* (or *ursprüngliche Stamm der Menschengattung*).[35] In Kant's view, "there are no different species *[Arten]* of human beings."[36] Even the present characteristics of white people, he insisted, had developed from the original predispositions in the *Stammgattung.*[37] This, for sure, was a considerably progressive position at the time, given that many of Kant's contemporaries considered nonwhite races not only inferior, but also unrelated and different in kind from white Europeans. On the other hand, numerous statements throughout Kant's work reflect his European privilege and racial bias. He speculated, for example, that although the particular traits and racial characteristics of the original *Stammgattung* are forever lost in time, it likely featured "whites with brunette color."[38]

By the time Kant began writing his *Third Critique* in the mid-1780s, he had already grown dissatisfied with his own theory. The problem was that despite his modifications, Kant's version of epigenesis remained too close to Haller's generic preformationism. Sure, the natural predispositions *(Anlagen)* introduced in Kant's model provided considerable flexibility for each phenotype to adapt to its specific environment. Still, there was no denying that the *Anlagen,* too, were ultimately preprogrammed and functioned only mechanically, not organically. Kant's distinction between *Keime* and *Anlagen,* in other words, did not change the fact that *both* germs and predispositions lay preformed within the organism. As with all theories of preformationism, the strength of this theory was that it only required a mechanical force of some kind or another to explain organic causation. Hence there was no need to introduce any other (vital) force into the mix (such as entelechy or divine intervention), which Kant deemed impermissible because it amounted to pure metaphysical speculation.

But there was one fatal weakness to Kant's theory: it did not correspond to reality. Scientific experiments at the time clearly showed that the development of life included non-mechanical forces and causation. Every child knew that if a worm is cut into half, it grows back some of its missing tissue. Living matter, unlike inert matter or man-made machinery, both self-constructs and self-maintains all its vital functions. This means that every part of an organism contributes to the creation of the whole, while, at the same time, the whole also informs and maintains the function of each of its constitutive parts. This causal recursivity, Kant realized, was inexplicable in terms of linear causality—that is, in terms of the mechanical forces of attraction and repulsion that defined both Newtonian physics and Kant's own view of the laws of nature. Unlike mechanical causation, which consists of a series of linear events that lead from A to B to C and so on, the recursive causation evident in living matter coils back on itself such that the "final" product retroactively affects and sustains its own "origins." A causes B, B causes C, and C closes the loop by interacting with and modifying A.

This recursive loop of auto-affection indicates that living organisms are both the means and the end, both cause and effect of their own being. How could this be? Besides mechanical forces, the only other possible explanation to account for this recursive self-organization and self-repairing ability of living matter, Kant reasoned, was to assume intentionality—that is, the presence of a willful agent or some other teleological force able to control the assembly and maintenance of heterogeneous parts into a coherent whole. This, however, could not be right either, in Kant's view, because free will was linked to rationality and thus unique to human beings. Unfamiliar with twentieth-century systems theory and the concept of emergence that help explain the genesis and development of life, Kant remained stuck between a mechanical and a teleological account, neither of which he thought could possibly be true.

Since the self-affective organization and behavior of living matter was irreducible to the preformationist logic of linear cause and effect, Kant started using the term "epigenesis" rather than "generic preformationism" to describe his own position, most notably in the second edition of his *CPR* from 1787. It would be a serious mistake, however, to interpret this terminological shift in the mid-1780s as evidence that Kant had completely abandoned preformationism in favor of epigenesis. John Zammito is certainly right to emphasize "Kant's persistent ambivalence towards epigenesis," although he tends to downplay the fact that Kant remained just as ambiguous vis-à-vis preformationist theories.[39] The most likely explanation, in my view, is that Kant simply switched labels to emphasize what he had come to recognize as the distinctive characteristic of living organisms—namely, their "self-organizing" nature.[40] This meant that all living matter "must be related to itself reciprocally as both cause and effect," as he put it in the *Third Critique* from 1789.[41] By that time, Kant had finally found—or thought he had found—what he had been seeking for more than a decade, namely, a biological theory able to reconcile the mechanical laws of the physical universe with the seemingly teleological behavior observable in embryology and living organisms. That theory was Johann Friedrich Blumenbach's notion of the "generative drive" (*nisus formativus* or *Bildungstrieb*).

Blumenbach had been an extraordinary professor of medicine in Göttingen before taking over Wolff's former chair at the school of physiology in 1789, a position that became available after Wolff had moved to St. Petersburg, Russia. Prior to the publication of his seminal text "On the Formative Drive" *(Über den Bildungstrieb und das Zeugungsgeschäfte)* in 1781, Blumenbach had actually been a proponent of preformationism and sympathetic to Haller's germ theory. During his studies on the regenerative power of the hydra (water polyp), however, Blumenbach became increasingly convinced that

> there exists in all living creatures, from men to maggots and from cedar trees to mold, a particular inborn, life-long active drive *[Trieb]*. This drive initially bestows on creatures their form, then preserves it, and, if they become injured, where possible restores their form. This is a drive (or tendency or effort, however you wish to call it) that is completely different from the common features of the body generally; it is also completely different from the other special forces *[Kräften]* of organized bodies in particular. It shows itself to be one of the first causes of all generation, nutrition, and reproduction. In order to avoid all misunderstanding and to distinguish it from all the other natural powers, I give it the name of *Bildungstrieb (Nisus formativus)*.[42]

With Blumenbach's notion of the *Bildungstrieb*, epigenetic theory at the end of the eighteenth century prepares the way for the vitalist philosophy of life *(Lebensphilosophie)* popularized by Nietzsche and Bergson a century later. The *Bildungstrieb*, according to Blumenbach, does not just trigger the development of some preexisting organic forms or unleash the God-given potential of some matter to become alive, as the preformationists had argued. Blumenbach's drive genuinely *created*, rather than merely unfolded, life from matter. An independent vital agency, the *Bildungstrieb* was "one of the first causes of all generation, nutrition, and reproduction" and "initially bestows on creatures their form."[43]

To this day, however, it remains unclear how unique Blumenbach's theory actually was. For one, his crucial distinction between the formative *power (Bildungskraft)* that inheres in *all* matter and the formative *drive (Bildungstrieb)* that applies only to *living* matter is clearly reminiscent of Buffon's earlier distinction between the basic gravitational versus the truly animating or "penetrating" forces of matter. In Germany, moreover, the young and ambitious Caspar Friedrich Wolff had already introduced another, yet similar theory in his dissertation from 1759 to account for the successive generation of organic forms. In particular, Wolff referred to "a principle of generation, or essential force *[vis essentialis]* by whose agency all things are affected."[44] This was not the kind of mechanical force Haller and the preformationists talked about. Instead, Wolff emphasized that the *vis essentialis* had genuine generative powers of its own. And whereas Buffon defined his "penetrating force" as "a physical property of matter," the *vis essentialis,* according to the young Wolff, was a distinct and separate force *external to matter.* This essential force provided the vital impetus necessary to organize previously inert, dead matter into organic, living forms.

By all accounts, it appears that Blumenbach's *Bildungstrieb* was strikingly similar to Wolff's *vis essentialis.* Yet both authors vehemently rejected this comparison at the time. Instead, they emphasized the unique character of their respective theories and accused each other of "obscurantism" and empty rhetoric bereft of explanatory power and scientific rigor. According to Blumenbach, Wolff's claim that the *vis essential* was external to matter meant it was akin to a living soul superimposed on matter, leading to a kind of vitalism Blumenbach considered unacceptable. Wolff, on the other hand, denounced Blumenbach's claim that the *Bildungstrieb,* though immanent to matter, was irreducible to it as a contradiction in terms that rendered this drive supernatural in origin and scientifically inexplicable.[45] As Wolff put it in 1789, the vis *essentialis* "does not produce the various parts of the body through itself and according to its nature (as in the *Bildungstrieb*) but with the aid of numerable other concurring causes; and what the essential force does on its own is always a simple effect, as attraction, or repulsion, worlds apart from the formation of organic bodies."[46] As this quote demonstrates, Wolff wanted to have it both ways. Although he criticized Buffon's "penetrating force" for being too mechanistic, he himself insisted on the "simple" mechanical nature of his own *vis essentialis,* which he described in classic Newtonian terms of attraction and repulsion so as to distinguish it from the alleged vitalism of Blumenbach's *Bildungstrieb.* In hindsight, it seems deeply ironic that each scientist denounced the other's theory as vitalist given that both are known today precisely *because of* rather than *despite* their contribution to the modern history of vitalist philosophy.

Kant, of course, was just as opposed to vitalism as were Blumenbach and Wolff. A good example is Kant's critical review of Johann Gottfried

Herder's *Ideen zur Philosophie zur Geschichte der Menschen* from 1785. A former student of Kant's, Herder had alienated his mentor with this speculative theory of life inspired by Wolff's writings. Kant explicitly rejected Herder's notion of an "inner life principle" as unscientific metaphysics. Instead, he kept promoting his own version of "epigenesis" (aka "generic preformationism") and suggested that Herder reformulate his theory in terms of the "germs or natural predispositions" (i.e., the *Keime* or *Anlagen*) that provide the "bare, not further explicable limitations on [the] self-structuring capacity" of life.[47] This last line about the "not further explicable" capacity of life to self-organize is crucial. There was simply no way, Kant argued, for rational thought to explain the genesis of living matter. The only way to deal with the problem was to accept this fact and stop speculating about a phenomenon that defies reason.

It was precisely this acknowledgment of our ignorance about the last principles of life that endeared Blumenbach to Kant. In the second edition of his book in 1789, Blumenbach explicitly refused to speculate any further about the origin and ultimate cause behind the *Bildungstrieb*: "I hope it will be superfluous to remind most readers," Blumenbach wrote, "that the word *Bildungstrieb*, like the words attraction, gravity, etc., should serve, no more and no less, to signify a power whose constant effect is recognized from experience and whose cause, like the causes of the aforementioned and the commonly recognized natural powers, is for us a *qualitas occulta*."[48] We have no idea from whence the *Bildungstrieb*—or gravity, for that matter—comes, Blumenbach readily admitted, meaning that the *ultimate* cause behind the emergence of life remains unknown and inexplicable to science.

It was this agnosticism that impressed Kant, because he took it to confirm his rejection of hylozoism. Thus the *Bildungstrieb* came to represent the middle way between preformationism and epigenesis Kant had sought for decades. Although Blumenbach's and Wolff's theories were closely related and offered similar solutions to the problem, Kant publicly embraced the former and rejected the latter, most likely because he was already biased against Wolff's theory due to its negative influence on Herder.[49] In a letter to Blumenbach in 1790, Kant wrote, "Your works have taught me a great many things; indeed your recent unification of the two principles, namely the physico-mechanical and the teleological—which everyone had otherwise thought to be incompatible—has a very close relation to the ideas that currently occupy me but which require just the sort of factual confirmation that you provide."[50]

In his *Third Critique,* Kant further commented on Blumenbach as follows:

> He begins all physical explanation of these formations with organized matter. For he rightly declares it to be contrary to reason that raw matter should originally have formed itself in accordance with mechanical laws, that life should have arisen from the nature of the lifeless, and that matter should have been able to assemble itself into the form of a self-preserving purposiveness by itself; at the same time, however, he leaves natural mechanism an indeterminable but at the same time also unmistakable role under this inscrutable principle of an original organization, on account of which he calls the faculty in the matter in an organized body (in distinction from the merely mechanical formative power that is present in all matter) a formative drive.[51]

As this passage makes clear, and as Robert Richards has argued at length, Kant *misunderstood* Blumenbach. According to Kant, Blumenbach argued that organized matter was indeed a primary substance always already present in nature, when, in fact, Blumenbach clearly stated that the *Bildungstrieb* works on and causes inert matter to become alive in the first place.[52] What remained inexplicable, for Blumenbach, was the exact principle behind this drive and its ability to vitalize matter, whereas, for Kant, it was the givenness of life itself—the fact that life, as he thought, was always already present in nature to begin with—that cannot be explained. Kant was prone to this misunderstanding because, as noted above, he was strictly opposed to hylozoism—the idea that life originally emerged from inanimate matter.

Kant and Blumenbach publicly praised each other's work for sure. Yet it seems more likely that neither of them fully understood what the other was saying. Despite this misunderstanding—or rather because of it—Blumenbach's *Bildungstrieb* provided the conceptual bridge Kant thought he needed to explain the "self-organizing" ability of life with reference to both mechanical and teleological forces. With the help of the *Bildungstrieb,* Kant felt comfortable to explain why the various parts of "a natural product" *(Naturprodukt)* can "be combined into a whole by being reciprocally the cause and effect of their form," because their development is guided by "a self-propagating formative power, which cannot be explained through the capacity for movement alone (that is, mechanism)," but must always already be at work in living matter.[53]

This explanation, however, is tautological. It begs the question it purports to answer. To explain the self-organizing ability of living beings in terms of their inherent ability to self-organize is hardly an explanation at all. The same circular logic, of course, haunts Kant's claim about the "epigenesis of pure reason," as Friedrich Nietzsche bluntly pointed out a century later. Kant's philosophy, Nietzsche argued, did little more than explain man's ability to reason simply "by virtue of a faculty"—or, as I would translate Nietzsche's original phrase, "by means of a means" *(vermöge eines Vermögens)*—and hence not at all. As Nietzsche put it, we might just as well say that opium makes us fall asleep because of its inherent power—the *virtus dormitiva*—to cause sleep.[54] The same tautology, indeed, characterizes Kant's account of living matter in terms of the *Bildungstrieb.* Note that this tautology did not exist in Blumenbach's account, because he, unlike Kant, considered the teleological *Bildungstrieb* a real, ontological force "whose undeniable existence and extensive effects are apparent throughout the whole of nature and revealed by experience."[55] While the ultimate cause behind the existence of this drive remained unknown, its material effects—namely, to create life from inert matter—can nonetheless be studied and systematized like that of gravity or any other force in nature. Kant, on the other hand, understood the *Bildungstrieb* not in ontological, but solely in epistemological terms. For him, it was a heuristic device necessary to explain a phenomenon that could not be explained otherwise. The tautological reference to the *Bildungstrieb,* in other words, allowed Kant to keep thinking of life "as if" it were teleological without, however, taking a stance on whether this presupposition was actually instantiated in reality.

In the *Third Critique,* Kant referred to this supposition as a "regulative judgment" or the "regulative use" of the concept of finality. Unlike "constitutive judgments," which, by definition, are constitutive of the phenomena we actually perceive, regulative judgments may—or may not—pertain to reality. As Kant saw it, we simply have no choice but to think of "organized entities" in teleological terms as purposive beings even if we suspect or have reason to believe they are not. Although the regulative concept of finality may not correspond to reality, it must be presupposed to make sense of—and come to terms with—the phenomenon of life in the first place. Otherwise, we are at our wit's end and cannot proceed to think about life at all. This is why regulative judgments play such an important role in Kant's transcendental philosophy: they strike a careful balance between epistemological necessity (to think "as if" regulative judgments were true) and ontological agnosticism (the admission that such judgments may not pertain to reality).

Kant's balanced approach, however, was soon abandoned by nineteenth-century biologists and philosophers alike. The German idealists and *Naturphilosophen* (e.g., Fichte, Schelling, Hegel) essentially collapsed epistemology and ontology into one and equated transcendental subjectivity with empirical subjectivity. Thus, the moment of apperception (the Kantian "I think") no longer served as a mere logical presupposition *for making sense of the world,* but was seen instead as *constitutive for being in the world.*[56] Human subjectivity—a purely transcendental concept in Kant's critique—now emerges as a self-generating and self-replicating substance in German philosophical discourse. A similar substitution of epistemological principles with ontological principles occurred in early nineteenth-century German biology. Kant used regulative concepts—like the notion that living organisms were purposive entities—as a necessary heuristic for further inquiry. The scientific community, on the other hand, had no qualms about making constitutive claims concerning living nature. Eager to prove the proper scientific status of their nascent discipline, German biologists (e.g., Reinhold Treveranus, Johannes Müller, Karl Ernst von Baer) both organicized and ontologized the *Bildungstrieb.* Unlike Kant, they were convinced that teleological principles did, in fact, govern the development of life, and they were determined to verify the *Bildungstrieb* empirically. In so doing, German biologists thought they followed Kant when, in fact, they followed Blumenbach.

The shift from the epistemological to the ontological realm was crucial for the rise and recognition of modern biology as proper science. As Philippe Huneman points out, "Kant's metaphysical point—that is, that the autonomy of the living realm is epistemological rather than ontological—is no longer salient in this perspective."[57] The alternative would have been for biologists to proceed under the crippling hypothesis that none of their conclusions might pertain to reality, that nothing they said was necessarily true, and that their entire experimental logic was dictated not by living nature, but by the insurmountable limits of human reason. Trying to establish a new scientific discipline on the basis of these premises would have been difficult, maybe even impossible, to do. Modern science remains open to falsification, for sure, but this process hinges primarily on concrete empirical evidence based on scientific data, not abstract philosophical reflections about methodological principles.

Emergent Life (Autopoiesis and Cognition)

There can be little doubt that Kant's philosophical reflections on life deeply influenced biological research in Germany during the first half of the nineteenth century. In fact, Timothy Lenoir went so far as to argue that Kant's attempt to reconcile teleological and mechanistic principles of life gave rise to a fifty-year-long, unified research project in German biology that Lenoir called "teleomechanism" or "vital materialism."[58] Although Lenoir's book was initially praised by scholars from across the disciplines, some historians of science have since come to criticize his claim about Kant's influence as oversimplified and far too generalized an account.[59] At least some of that critique, I would argue, is due to the historical misunderstanding between Blumenbach and Kant. As mentioned above, this misunderstanding led nineteenth-century biologists like Treveranus and von Baer to believe they followed Kant even though they did not, which, in turn, may have led Lenoir to overestimate Kant's contribution to early German biology.

Be that as it may, Lenoir's most significant contribution, in my view, has actually less to do with Kant than with the history of vital materialism and its relation to science. Lenoir clearly showed that vitalist theories of life were not just present, but hegemonic in pre-Darwinian German biology, which offered a wide spectrum of different versions of teleomechanism. Many of these versions were vitalist in nature even if they did not use or advocate that specific term. The crucial point is that most biologists at the time acknowledged the purposive behavior of living organisms and cautioned against an overly reductive account of biological processes solely in terms of physical laws. Teleological explanations of life, however, lost credibility after Darwin's introduction of natural selection in 1859 and the great synthesis of Darwin and Mendel in the 1930s. After that, the life sciences began to reject any reference to teleological principles as impermissible metaphysics. Particularly since the discovery of the double helix in the 1950s, the erstwhile modesty that had characterized Kant's regulative framework for explaining the phenomenon of life was replaced by an increasingly boastful rhetoric about the power of genetic research to decode, and possibly rewrite, "the book of life" in all its complexity, from the micro level of "the selfish gene" (Richard Dawkins) to the macro level of "sociobiological" structures and human cultural evolution (E. O. Wilson) *without any reference to teleology or purpose at all.*

If we accept this neo-Darwinian framework, there is indeed little reason to return to Kant. Indeed, most evolutionary theorists today emphasize the scientific *limitations* of Kant's transcendentalism, not its insights. Since biology is a natural science, this means, first of all, that the ontological realm comes back into play, as mentioned above. According to Darwinian theory, the epistemological fit between (perceiving) subject and (perceived) object results from the evolutionary adaptation of our species to its environment over hundreds of thousands of years: "Subjective epistemological structures fit/correspond to the world," Gerhard Vollmer declares in explicit contradiction to Kant, "because they have evolved during evolution precisely by adapting themselves to this world."[60] In other words, there is no "pure reason a priori," because human reason and knowledge are both based on experience—albeit on the level of the genotype, not the phenotype: "Everything comes from experience," Jacques Monod insists, "yet not from ongoing current experience, reiterated by each individual with each new generation, but instead, from the experience accumulated by the entire ancestry of the species over the course of its evolution."[61] Which is to say that Kant's transcendental reason is empirical through and through, because it results from the history of ancestral experience and human evolution shaped by natural selection.

Second, the fact that human beings share most of their evolutionary history with other species also suggests that even our most distinctive traits—such as rationality, language, culture, cognition—exist across a wide-ranging spectrum and are shared, to varying degrees, with other animals. It follows that Kant's categorical distinction between humans and all other forms of life, including animals, is untenable. "There is no longer any good reason to take it for granted that the theoretical, ethical, and political question of the subject is automatically coterminous with the species distinction between *Homo sapiens* and everything else," Cary Wolfe pointed out in 2003. He argues that numerous empirical studies have shown "that the traditionally distinctive marks of the human (first it was the possession of a soul, then 'reason,' then tool use, then tool *making,* then altruism, then language, then the production of linguistic novelty, and so on) flourish quite reliably beyond the species barrier."[62]

I believe these specific objections to Kant are valid. In his defense, one might want to point out that Kant is hardly the only Western philosopher guilty of anthropocentrism, and that, moreover, he lived and wrote prior to Darwin. More important, however, is to recognize that the hegemony of the neo-Darwinian view in contemporary biology has served to obscure the methodological and epistemological questions Kant raised. Over the last decade in particular, scholars like Evan Thompson, Terrence Deacon, and Jennifer Mensch have argued that Kant's insights regarding the self-organizing nature of living matter, as well as the epistemological conclusions he drew from these insights, remain as relevant today as they were some 250 years ago.[63] To understand why, we need to remember that Darwin never answered his own question as to the "origin" of the species or life itself. Instead, Darwin's theory of natural selection presupposes the existence of life without trying to explain how it came about in the first place. The neo-Darwinian focus rests exclusively on the survival and reproduction of (phylogenetic and ontogenetic) individuals as opposed to their original formation.

That, however, was Kant's main question. Given his lack of resources, the only way for him to explain the genesis of life was to bracket the ontological dimension and declare the purposive behavior of living matter a mere regulative principle of reason. Today, we can use complexity and systems theory to fill in the blanks and augment his philosophical account of life. Reading Kant's *Third Critique* alongside autopoietic theory, Thompson, for example, claims that autopoiesis provides a genuine biological explanation for Kant's two central claims: first, that living organisms differ from artificial machines because they are self-generating (autopoietic) systems, and second, that their purposive behavior can be understood in terms of their ability to make sense of their environment, which allows them to adapt their internal structure in response to external stimuli. In short, "What Kant recognized as a distinguishing characteristic of organic beings—that they are unities rather than mere aggregates—finds its minimal expression in a living cell."[64]

While there is general agreement among philosophers of science about the prescience of Kant's philosophy with regard to autopoietic theory, there still remains considerable tension about how exactly to conceptualize this relation and how best to describe the phenomenon of life in autopoietic terms. The ongoing debate about the relation between autopoiesis and cognition is a good case in point. When Maturana and Varela first coined the term autopoiesis in the late 1970s, they claimed that autopoiesis and cognition are coeval, meaning that all autopoietic systems are also cognitive systems. Yet critics since then have argued that some simple chemical systems, though autopoietic, lack an internal metabolic network and hence cannot change their own structure in response to environmental stimuli. Lacking adaptivity, such autopoietic systems are clearly not cognitive, and therefore nonliving, according to Evan Thompson.[65] In his detailed discussion of the history of autopoietic theory, John Protevi makes a similar point. He demonstrates that the original concept of autopoiesis focused exclusively on how living systems maintain their homeostasis and organizational closure, but failed to consider how these systems interact with and continuously adapt to their particular environment. According to Protevi, this changed in the late 1980s, when Varela began to embrace a diachronic notion of autopoiesis to account for the production of new functional structures within the system in response to external perturbations.[66]

Bioaesthetics adopts this latter, broader definition of autopoiesis as a self-generating, living system engaged in interactive processes with its environment, which includes cognition or "sense-making," as Varela called it. But it also acknowledges the relevance of other theories that are difficult to reconcile with Varela's and Thompson's view. According to Deacon, for example, life emerges through the synergy of two dynamic chemical processes—namely, autocatalysis (i.e., a self-amplifying process of positive feedback loops among interacting molecules that continues to accelerate until it exhausts itself or meets external constraints), and the self-assembly or self-containment of such processes via the spontaneous emergence of some real physical boundary, like a membrane, that separates the resulting molecular structure from its environment and maintains its existence against the forces of entropy. The synergetic reciprocity of autocatalytic and self-containing processes, Deacon explains, "produces a special kind of emergent stability, unavailable to either process in isolation." The result is what he calls an "autogen," that is, a genuinely self-generating, self-repairing, and self-replicating system. Although Deacon considers the term autogen "closely related" to Maturana and Varela's term of autopoiesis, the two terms differ insofar as autopoiesis emphasizes the *process* of self-generation, whereas autogen emphasizes the *product* of such processes, namely, "some form of individuality" or "virtual self." This focus on the autogenic "self" explains why Deacon insists on a real physical boundary between living systems and their environment, because without this boundary, there is no "self," so to speak.[67] Thompson's theory, on the other hand, is based on autopoiesis, not autogens. His definition of life, therefore, focuses on the structural openness of living systems and their ability to adapt to their environment, which requires only a functional border and not a real physical boundary between the two.

The best way to illustrate this difference is to examine the question of whether viruses are alive. Since Thompson focuses primarily on autopoietic processes and the ongoing exchange between livings systems and their environments, he argues that viruses are neither alive nor autopoietic, because they have no metabolism of their own and do not produce their own nucleic acids, but instead use those of their host.[68] Deacon, on the other hand, claims viruses are alive because they are self-contained entities protected "by protein shells or more complex coverings."[69] Note that these opposite views are not mutually exclusive, but are based on different perspectives of the same phenomenon. If we see life as a metabolic process, viruses are not alive; but if we see life as a virtual self, then they are alive.

These opposing views of life also differ on the cognitive abilities they ascribe to autopoietic systems. I already mentioned that, according to Varela and Thompson, all autopoietic systems that interact with their environment should be recognized as cognitive systems.[70] Margulis and Sagan, in fact, take this view even one step further. Their crucial point is not only that all autopoietic systems are cognitive systems, but that "every organic being, every autopoietic cell is conscious."[71] This seems overstated.[72] Cognition is not the same as consciousness, and not all forms of consciousness include subjective experiences or self-awareness. Deacon, finally, recognizes basic autopoietic systems (or what he calls "autogens") as "a form of individuality," yet, at the same time, denies that these "individuals" have any kind of cognition at all, let alone consciousness or a genuine will of their own (or what he calls "ententionality"): "The exploration of non-linear and far-from-equilibrium dynamical systems," Deacon argues, "has not provided an account of the ententional properties that originally inspired the development of emergence theories." Autogens are "sentient," in his view, yet they lack emergent phenomena like "cognition," "consciousness," or "subjectivity" that characterize higher and more complex forms of life. Hence the "ententional properties" scientists commonly ascribe to autopoietic systems "must be treated as mere descriptive heuristics."[73]

I have engaged this complex debate at length in order to demonstrate, once again, how difficult it is for philosophers and scientists to come to terms with the phenomenon of life. Equally important, in my view, is the ongoing relevance of Kant's intellectual legacy in this debate. Deacon's understanding of the ententionality of basic autopoietic systems "as mere descriptive heuristics" clearly echoes Kant's view in his *Third Critique*, according to which the seemingly "purposive behavior" of basic organisms is but a regulative idea of human reason. At the same time, however, emergence and autopoietic theory clearly confirm what Kant ardently denied—namely, that life emerged from inert matter (i.e., hylozoism).[74] Kant, for sure, was ready to concede that living matter is self-*organizing*, but he refused to believe that it is self-*generating*.

Apart from the Kantian legacy, what matters most to bioaesthetics is that autopoietic theory challenges the neo-Darwinian dogma about the primacy of reproduction and natural selection for the evolution of life. Deacon insists that "natural selection emerges from the dynamics of reciprocally reinforcing self-organizing processes," not the other way around.[75] Life *emerged* from autopoietic systems; it did not *evolve* from natural selection, as orthodox Darwinians claim, because that process, obviously, requires something to be there first, *before* selection can take place. Despite their differences, the proponents of autopoiesis (e.g., Varela, Deacon, Thompson, Protevi, Margulis and Sagan) agree on this point, namely, that the neo-Darwinian emphasis of reproduction puts the cart before the horse. "To conceive of the first lifelike process," Deacon argues, "we must remind ourselves that it was not reproduced, it had no parents, and therefore it did not evolve. It emerged. It was fitted to its environment by chance alone, not by virtue of having been tested by natural selection."[76] Thompson concurs: "Reproduction presupposes autopoiesis, but autopoiesis does not necessarily entail reproduction, for a system can be self-producing according to autopoietic criteria without being capable of reproduction."[77]

As I stated in my Introduction, I think that bioaesthetics, unlike philosophy or the sciences, has no reason to enter the fray and should recuse itself from speculating about what came first, the chicken or the egg. Bioaesthetics can afford to remain agnostic about life's origin and first principles, because that debate is largely irrelevant to its particular objects of study. The sole relevance this debate holds for bioaesthetics, in my view, is the serious challenge it poses to the excessive reductionism of today's biologistic theories of human art and culture based on Wilson's consilience and neo-Darwinian orthodoxy. The "aesthetic faculty" that distinguishes human beings from other animals, Deacon rightly concludes, is an emergent "cognitive capacity that is far from genetically prefigured" and irreducible to the process of natural selection.[78]

Instead of neo-Darwinian theory, bioaesthetics embraces the autopoietic notion that all living organisms are cognitive systems able to enact their unique environment by making simple choices. Hence Varela insists again and again that "living is sense-making":[79] "Consequently, cognition is no longer seen as problem solving on the basis of representations; instead, cognition in its most encompassing sense consists in the enactment or bringing forth of a world by a viable history of structural coupling."[80] As Margulis and Sagan put it, "Life is distinguished not by its chemical constituents but by the behavior of its chemicals."[81] This behavior is *constrained* by the laws of physics and biological principles, but it is *not determined* by them.

I have argued throughout this chapter that Kant had already come to the same conclusion in his *Third Critique* with regard to the recursive, nonlinear causality at work in living matter. The question we need to engage now is why this matters to art and aesthetics. My main argument will be that aesthetic judgments in Kant's sense amount to a self-conscious reflection on the autopoietic nature of human cognition. Such judgments are paradigmatic of the "self-birth of reason," because they suspend the normative framework of conceptual thought that accompanies and informs everyday human behavior. In so doing, aesthetic judgments not only return our mind to the ongoing process of making sense that characterizes living systems, but also force us to come to terms with the peculiar nature of being human. This means that "our attention to beautiful objects and events, and our experience of aesthetic enjoyment, may coherently be understood as the results of a biological need to locate certain types of information in our environments, as a supplement to genetic information, for the purpose of constructing and maintaining our own order," as Francis Steen recently put it. Without reference to Kant, Steen's cognitive theory of aesthetics summarizes what I believe Kant had in mind, namely, that "the very design of aesthetics and imaginative play is to explore a vast phase space of human action, much of which has not been realized and thus cannot have been acted on by natural selection."[82] This self-transformative exploration of our mind, I argue, is the biological origin of the *sensus communis aestheticus* that, according to Kant, is essential to being human.[83]

Sensus Communis Aestheticus

Following the table of the categories outlined in the *Critique of Pure Reason*, Kant's *Third Critique* defines aesthetic judgments (of beauty) in terms of quality, quantity, relation, and modality. In summary, a beautiful object, according to Kant, arouses feelings of "satisfaction or dissatisfaction without any interest" (*CJ,* 96; *KdU,* 288, A17/B17); it "pleases universally without a concept" (*CJ,* 104; *KdU,* 298, A32/B32); it displays "purposiveness . . . without representation of an end" (*CJ,* 120; *KdU,* 319, A61/B62); and it "is cognized without a concept as the object of a necessary satisfaction" (*CJ,* 124; *KdU,* 324, A68/B69). Each of these qualifications in *Third Critique* has been submitted to endless commentary and critique, the detailed analysis of which far exceeds both my ability and my ambition in this book. I shall focus my attention instead on two sentences only, beginning with the seminal question that introduces paragraph 9: "Investigation of the question: whether in the judgment of taste the feeling of pleasure precedes the judging of the object or the latter precedes the former."[84] This question about the temporal structure of aesthetic judgments is crucial to Kant's aesthetics and remains a major point of contention among scholars even today. Kant himself emphasized that it is "the key to the critique of judgment, and hence worthy of full attention."[85] Yet Kant had a difficult time trying to answer his own question. In some passages, he clearly implies that pleasure precedes judgment, whereas in others, he argues the reverse. For if our feeling of pleasure were indeed primary and hence "merely subjective" *(bloß subjectiv),* it would be neither disinterested nor communicable to others, in which case his crucial distinction between genuine aesthetic judgments and mere sensual judgments *(bloßes Sinnenurteil)* becomes null and void. Hence our feeling of pleasure must be the effect of aesthetic judgments, not their cause: "Thus it is the universal capacity for the communication of the state of mind in the given representation which, as the subjective condition of the judgment of taste, must serve as its ground and have the pleasure in the object as a consequence."[86] In this and similar passages throughout the *Third Critique,* it is the communicability of our mental state affected by beauty that provides the foundation for a reflective judgment, which, in turn, gives rise to the feeling of pleasure.

Yet this temporal logic, too, causes serious problems in Kant's aesthetics. For if reflection precedes our feeling of pleasure, it remains unclear *what* exactly the mind is supposed to reflect about, as Paul Guyer points out.[87] This contradiction, Guyer argues, can only be avoided if we accept that aesthetic judgments of taste entail not just one, but two separate kinds of judgments: first, a "simple reflection" about an object produces a feeling of pleasure or unpleasure before giving way to a second, genuinely "reflective judgment" about this feeling caused by the first judgment. Guyer's two-phase model, to be sure, dissolves the contradiction at the heart of Kant's theory of aesthetic judgments and, quite literally, sets the matter straight. On the other hand, however, his linear logic abandons what appears to be Kant's main concern—namely, to avoid prioritizing either one of the two constitutive moments of aesthetic experience. Kant's going back and forth on this central question suggests instead that feeling and reflection are synchronous and structurally codependent features of aesthetic judgments.

In order to salvage this insight—that is, the recursivity of feeling and reflection as mutually constitutive moments of aesthetic judgments—Peter Gilgen interprets Kant's formulations in system-theoretical terms as a description of what Niklas Luhmann calls "second-order observations." Unlike first-order observations that focus on a given object, second-order observations reflect on the act of observation itself and thus turn first-order observations into their object of analysis: "The judgment of taste," Gilgen concludes, "is nothing but a second-order observation that reflects upon itself."[88] Unlike Guyer's linear model, which divides aesthetic reflection into two consecutive moments and two separate kinds of judgments, Gilgen's interpretation retains its genuine circular structure, because second-order observations differ from first-order observations solely with regard to their object, but not with regard to their structure. In both first- and second-order observations, there always exists only one kind of observation, not two. Since first-order observations serve as the constitutive objects for second-order observations, the two are fused together and cannot be divvied apart into two separate processes. This enables second-order observations to observe the act of observation.[89] Aesthetic judgments, Gilgen argues, are self-reflexive in the same way: they reflect on themselves as objects, and these objects emerge—and can emerge *only*—during the act of reflection itself.

Gilgen's reformulation of Kant's aesthetics in terms of Luhmann's systems theory helps clarify the crucial point that aesthetic judgments exhibit the same recursive (autopoietic) structure Kant also considers the key characteristic of living organisms and the "self-birth" of human reason. To appreciate the significance of this point, we must recall that aesthetic judgments are reflective judgments and thus differ from determining (cognitive) judgments. The latter, Kant argued, subsume the particular under the universal, meaning they categorize a particular phenomenon according to predetermined, general rules. Reflective judgments, by contrast, only have access to the particular and must find—and not just apply—the general rule or concept that accords with this particular.[90] The particular characteristic of reflective *aesthetic* judgments, however, is that the general rule they try to find cannot be conceptualized at all. Aesthetic judgments, according to Kant, are based on the free play of the mind's faculties and remain irreducible to either rules or concepts, because every such "play" is unique and constitutes a singular occurrence expressed in that particular judgment. Hence Kant refers to the "exemplary" necessity of aesthetic judgments: it is "an example of a universal rule that one cannot produce."[91] Aesthetic judgments exemplify the necessity of a general rule that cannot be formulated in terms other than its exemplification. Aesthetic judgments are singularities. They are both cause and effect of their universal validity.

Let me clarify this paradoxical nature of aesthetic judgments with reference to another sentence—namely, the headline of paragraph 22, the final paragraph in Kant's analytic of the beautiful in the *Third Critique*. Since much of my argument hinges on the syntax and specific wording of Kant's original phrase, I shall quote this headline in German first, followed by its English translation:

> Die Notwendigkeit der allgemeinen Beistimmung, die in einem Geschmacksurteil gedacht wird, ist eine subjektive Notwendigkeit, die unter der Voraussetzung eines Gemeinsinns als objektiv vorgestellt wird. (*KdU*, A66/B67)

> The necessity of the universal assent that is thought in a judgment of taste is a subjective necessity, which is represented as objective under the presupposition of a common sense. (*CJ*, 123)

Though a fairly simple specimen by Kant's standards, this sentence is not easy to disentangle, both in terms of its syntax and with regard to its meaning. The main clause ("The necessity of the universal assent . . . is a subjective necessity") is supplemented by two relative clauses, the first of which modifies the genitive object ("universal assent, that is thought . . ."), while the second modifies the predicate noun ("subjective necessity, which is represented . . .") of the main clause. The meaning of the first relative clause does not pose any particular problems, except for using the term "judgment of taste," which in this case—though not always—is synonymous with genuine "aesthetic judgment."[92] Together, the main and relative clauses clarify, once again, that aesthetic judgments are characterized by a dual necessity: they are subjective (not objective), but also universal, because they presume that everybody—not just the individual person making the judgment—agrees with them, at least in principle.

There is considerable tension between these two requirements. A subjective judgment, one might say, is precisely that—subjective, meaning neither objective nor universal. Kant, however, differentiates between objectivity and universality precisely in order to distinguish aesthetic judgments from cognitive judgments based on objective concepts, on the one hand, and mere sensuous judgments based on individual taste, on the other. Since aesthetic judgments are non-conceptual, they clearly differ from the objectivity of cognitive judgments, so that problem is solved. But aesthetic judgments must also be universal to distinguish them from merely sensuous judgments *(bloßes Sinnenurteil)*. Both kinds of judgments are subjective (not objective, as are cognitive judgments), but aesthetic judgments require universal assent, whereas sensuous judgments do not. If they did not require universal assent, aesthetic judgments would be entirely arbitrary and merely reflect the speaker's personal interests and desires, in which case the crucial distinction between aesthetic and sensuous judgments would collapse. Kant realizes, of course, that we may—and often do—quarrel about personal likes and dislikes with regard to objects and their qualities (e.g., color, shape), including quarrels about whether a particular object is beautiful. But Kant's point is not that everybody does, in fact, agree with a particular judgment of beauty right away. That, precisely, is why the communicability of aesthetic judgments is so important, as we saw in paragraph 9. If we did not express our judgment or voice disagreement with the judgment of others, neither aesthetic experience nor aesthetic education would be possible at all, because it would prevent us from checking whether a person's judgment is genuinely aesthetic. Kant's main point, once again, is that any person who claims "this is beautiful" presupposes that others cannot but agree with this judgment *in principle*. Whether they actually *do* agree is a different matter and hinges on empirical questions that remain subject to debate, such as: Is the person's judgment disinterested, or does it reflect individual preferences? Is this judgment based on concepts or not?

The central argument of paragraph 22 is spelled out in the second relative clause of the headline. Its task is to mitigate the semiotic tension between the subjectivity of aesthetic judgments and their universal validity. At first sight, however, this phrase seems to make matters worse, because it links the subjectivity of aesthetic judgments not just to universality, but to objectivity as well. Kant now claims that the necessity of universal assent is a subjective necessity that "is represented as objective under the presupposition of a common sense." It seems contradictory—or, at the very least, counterintuitive—to represent something subjective ("a subjective necessity") as objective. The passive construction further acerbates the semantic obscurity of this phrase, because its grammatical subject is not specified. It remains unclear who, or what, is responsible for "representing" a subjective judgment as objective. The headline merely states that all this "is thought" and "is represented," but it does not specify by whom—a human individual, we must assume, yet one whose judgment is not merely subjective, but also universal. For whoever makes the judgment expresses a common feeling presumably shared by everybody else. Hence the real agent behind that which "is thought" and "is represented" in aesthetic judgments can be none other than the transcendental subject—that is, human reason itself.

My key point is that Kant, by means of his syntax, empowers reason to act. It is none other than reason itself that ensures that the subjective necessity of universal assent "is represented *[vorgestellt wird]* as objective under the presupposition of a common sense." Kant's use of the German verb *vorstellen* is particularly telling in this context. Although Guyer's translation of *vorstellen* as "to represent" is correct, the English version does not capture the nuanced meanings of the German original. *Vorstellen* literally means to put something in front of, or ahead of, something else.[93] This suggests that human reason does more than just "represent" a subjective judgment as objective. It would be more apt to say that reason literally enacts the objectivity of aesthetic judgments. The Grimm German dictionary from 1854 supports this reading. It states that the intransitive passive voice of *vorstellen* (i.e., *vorgestellt wird,* as used by Kant) is frequently used in philosophical texts to indicate "the *agency* of representation itself."[94] In this sense, we might say that the act of judging *creates* the very objectivity that the judgment, on its semantic level, claims to *re-present*. This recursive logic implies that the temporal lag between the creation and the representation of objectivity no longer exists once we reach the end of Kant's sentence. Put differently, paragraph 22 entails the same temporal paradox and autopoietic structure we already encountered in paragraph 6 regarding the simultaneity of feeling and reflection in aesthetic judgments. In both cases, it is impossible to say which part—a feeling or its reflection, a subjective judgment or its objective representation—comes first and conditions the other.

Let me support this conclusion with reference to the central concept of paragraph 22—namely, "common sense." The subjective necessity of universal assent, Kant states, "is represented as objective under the presupposition of a common sense." The philosophical history of the term "common sense" *(Gemeinsinn)* dates back to ancient Greece. In *De Anima,* Aristotle had argued that humans and animals must have a central perceptual capacity that monitors and unites the five regular senses. He called this higher-order capacity the *sensus communis.* The Greek tradition thus constitutes the first, *intrasubjective* meaning of the *sensus communis* as a (cognitive) meta-sense, one able to combine our ability to see, hear, smell, touch, and taste the world around us. Dating back to the ancient Romans and Stoic philosophy, the second meaning of the *sensus communis* is not intra-, but intersubjective. Also translated as "common sense," this intersubjective meaning of the *sensus communis* dominates early modern British and Scottish philosophy. It denotes a basic cognitive grasp of the world shared by all reasonable people and, in principle, by all members of the human species. This modern understanding of the *sensus communis* (as common sense) has since been defined in different, often contradictory terms, be it as inherently malevolent (e.g., Hobbes's *homo homini lupus est*), or as inherently benevolent (e.g., Montaigne's and Rousseau's noble savage), or as purely self-interested (e.g., Adam Smith's *homo oeconomicus*), et cetera. Although the connection is rarely made explicit, today's biologism, too, is clearly based on this inter-subjective meaning of the *sensus communis,* understood as the (genetically inherited) common sense that unites "the universal people" of humanity.[95]

Kant's *Third Critique* clearly refers to the modern, intersubjective (non-Aristotelian) tradition of the *sensus communis.* At the same time, however, Kant emphasizes that in contrast to the British understanding of "common sense," which is based on concepts, his own understanding of the *sensus communis* is based on the free play of the human faculties without concepts. Genuine aesthetic judgments, he argues, are possible only because we presume the existence of a universally shared "subjective principle, which determines what pleases or displeases only through feeling and not through concepts, but yet with universal validity."[96] Kant defines this principle as the *Gemeinsinn (sensus communis aestheticus),* which he explicitly distinguishes from the *gemeinen Verstande (sensus communis logicus)* used in the British tradition.[97] Aesthetic judgments, in other words, are based on the presupposition of the *sensus communis aestheticus* (as opposed to the *sensus communis logicus*). This presupposition alone ensures the universal validity and inter-subjective communicability of aesthetic judgments: "This indeterminate norm of a *sensus communis* is really presupposed by us: our presumption in making judgments of taste proves that."[98] Every time we venture to make an aesthetic judgment, Kant argues, we do, in fact, presume that something like a *sensus communis* exists, which is why we believe that others must agree with us in principle. Yet Kant explicitly refuses to speculate as to whether the *sensus communis* does, indeed, exist:

> Whether there is in fact such a *sensus communis*, as a constitutive principle of the possibility of experience, or whether a yet higher principle of reason only makes it into a regulative principle for us first to produce a common sense in ourselves for higher ends, thus whether taste is an original and natural faculty, or only the idea of one that is yet to be acquired and is artificial, so that a judgment of taste, with its expectation of a universal assent, is in fact only a demand of reason to produce such an unanimity in the manner of sensing, . . .—this we would not and cannot yet investigate here.[99]

Although the *sensus communis* constitutes the sine qua non of aesthetic judgments, and although we presume it exists every time we make an aesthetic judgment, we cannot know whether it *already* exists or whether it is just a demand of reason that it *should* exist. Hence all we can do is to *presume* it exists in order for judgments of taste to be universal and make sense. For this reason, the headline of paragraph 22 clearly states that aesthetic judgments, though subjective, are "*represented* as objective under the *presupposition* of a common sense." The *sensus communis,* Kant admits, "is a merely ideal norm" *(eine bloße idealische Norm),* the presupposition of which guarantees the universal validity required by aesthetic judgments.[100] In other words, these judgments are informed by and based on the presupposition of an ideal norm (i.e., the *sensus communis* resulting from the free play of the faculties) that can be specified solely in the utterance of such judgments.

Much like Kant remained agnostic about the ontology of the *Bildungstrieb,* he remained agnostic as to whether the *sensus communis* is a real human faculty or merely a regulative principle of human reason. Given the autopoietic structure of aesthetic judgments, our best option to solve this dilemma might be to reject this binary and fold both propositions into one. This means that the *sensus communis* is enacted and becomes real in and through aesthetic judgments. Its normative ideal cannot be determined conceptually, because it only emerges in exemplary fashion each and every time we make a genuine judgment of taste.

At this point, the decisive difference between today's biologist notion of "human nature" and Kant's notion of *sensus communis* should become clear. Neo-Darwinian scholars believe it is "possible to define human nature," the existence of which, they claim, "is no longer in doubt."[101] By defining human nature in terms of scientific concepts, biologism assimilates aesthetic judgments to cognitive judgments. It collapses the distinction between the *sensus communis* and common sense, which is as grave an error as to reduce the mind to the brain or biology to physics. Kant noticed the difference. For him, our sense of human nature emerges from the free play of the faculties. It can only be articulated qua aesthetic judgments, because each and every (successful) judgment exemplifies the ideal norm, the *sensus communis*, presumably shared by human subjects. Aesthetic judgments do not exemplify preexisting norms or scientific concepts. Instead, they literally enact the coming into being of such norms and concepts. Biologist discourse, on the other hand, flaunts the logical deduction of aesthetic judgments from first principles. Its revolutionary claims about the nature of art and aesthetics are predicated on—and hence do nothing but reconfirm—the ontology of "human nature" on which they are based in the first place.

Let me repeat, once again, that I do not object to the scientific definition of human nature insofar as it guides biological research and practice. My problem is not with biology, but with biologism—understood as the effort to reduce everything under the sun, including art and human culture, to evolutionary principles defined in neo-Darwinian terms. The previous discussion of Kant's philosophy, I hope, has clarified why science cannot possibly solve the metaphysical and methodological problems it encounters but is forced to ignore for the sake of producing empirical knowledge. If we systematically deduce the "nature" of art and culture from the "genetic blueprint" and "mental architecture" common to us, "the universal people," then genuine aesthetic experience in the Kantian sense simply ceases to exist. Because deductive reasoning yields determinate judgments that foreclose the free play of the faculties, as Kant would say—which, after all, is the raison d'être of art and aesthetic experience in the first place. Their cognitive potential, if we might call it that, is due precisely to the unpredictability of the *sensus communis* instantiated in each particular judgment. Kant considers human nature a normative ideal that art and the aesthetic faculty inspire within us; it is not a biological–ontological substrate one always already possesses as a member of the species. Once conceptualized in purely scientific terms, the *sensus communis aestheticus* as the ideal norm of human nature is irretrievably lost. It is, above all, this loss of aesthetic sense that characterizes neo-Darwinian studies of art and culture today.

Biologism and the proclaimed "revolution in the humanities," in my view, results in a big step backward rather than forward with regard to interdisciplinary research and the hope for better collaboration between the two cultures. It blatantly ignores the methodological foundation and disciplinary sovereignty of aesthetic discourse and its rich intellectual history since 1750. The same is true of other current philosophical strands, such as object-oriented ontology and speculative realism, both of which claim to move beyond Kant's correlationism but end up falling way behind it. The basic problem is that by choosing to ignore Kant, critics cannot but ignore the important epistemological questions originally raised by Kant and elaborated today in great detail by systems theory and cognitive science—questions about the genesis and development of life, about the intentionality and purposive behavior of living organisms, and about the antinomies of human reason. In the next chapter, I want to demonstrate the continued relevance of these questions and Kant's legacy in the context of Marxist philosophy and its relation to modern science.

2 MARXISM AND BIOLOGY

Meaning cannot be found in the movement of molecules.

RICHARD LEWONTIN

Marx and Darwin

"The left needs a new paradigm," Peter Singer argued in his 1999 manifesto *A Darwinian Left*. That new paradigm should be evolutionary theory to help correct the mistaken belief "that there is no fixed human nature," a belief that, according to Singer, continues to shape the left's knee-jerk rejection of Darwinism.[1] Many on the left, Singer claimed, profoundly misunderstand Darwinian theory: they identify it exclusively with the ruthless "struggle for existence" and "the survival of the fittest" elaborated by Herbert Spencer in the late nineteenth century, which led to the rise of twentieth-century totalitarianism and its racist politics of "social Darwinism." Singer, by contrast, rightly pointed out that Darwin's evolutionary theory lends itself just as well to the social study of cooperative behavior among humans. Cooperation *can* increase a species' chance for survival and reproduction, as numerous biologists have pointed out since the 1960s.[2] Singer also claimed that due to their misunderstanding of Darwin's theory, the majority of Marxist thinkers, from Engels and Lenin to Lysenko and official Soviet anthropology, falsely assumed that human cultural evolution is more or less *unconstrained* by human nature and hence cannot—indeed, it must not—be understood in biological terms lest it "dashed the left's Great Dream: The Perfectibility of Man."[3] According to Singer, however, there are many "common human tendencies that transcend cultural variation," such as incest taboo along with birth, marriage, and burial rituals.[4] Together, these universally shared traditions across different cultures provide empirical evidence for the existence of our common human nature.

In making this argument, Singer explicitly aligned himself with—and drew scientific support from—a vocal group of neo-Darwinians and evolutionary psychologists (e.g., Richard Dawkins, Daniel Dennett, Steven Pinker, Leda Cosmides, and John Tooby), who likewise argue that human behavior is severely constrained by our genetically determined "mental architecture" inherited some two hundred thousand years ago from our ancestors during the Pleistocene era. Given these evolutionary constraints, Singer cautions that "those seeking to reshape society must understand the tendencies inherent in human beings," which means that the leftist "dream of perfectibility should be put behind us."[5]

There are numerous shortcomings in Singer's book, as his critics were quick to point out.[6] Marx did, after all, develop a cultural anthropology of sorts in his early writings. The *Economic and Philosophic Manuscripts of 1844,* for example, conceptualize "the essence of man" in species terms as a laboring animal. The human being, Marx argued, can realize itself *as human* only in and through the deliberate, goal-oriented action called labor, and this labor, Marx believed, is "an exclusively human characteristic" irreducible to animal behavior.[7] He developed a similar argument in his *Theses on Feuerbach* a year later.[8] It bears repeating that much of the century-old confusion surrounding Marxism's alleged rejection of human nature is due to the historical influence of Engels, Lenin, and Stalin in twentieth-century Marxism. In many of their writings, Marx's followers effectively severed culture from nature at the expense of Marx's own and far more subtle reflections on human history. Engels, for example, claimed that "Marx's inquiry begins precisely where Darwin's inquiry ends."[9] Hence he believed that just like "Darwin discovered the law of development of organic nature, so Marx discovered the law of development of human history."[10] Engel's formulations are unfortunate, to say the least. They imply that natural evolution and human history are two separate phases of development when, in fact, both unfold simultaneously and constantly interact with each other across time and space.

The young Marx, on the other hand, more than once acknowledged the constitutive intertwinement of natural and cultural evolution throughout human history. Far from being anti-Darwinian, Marx merely considered human nature more adaptable and less constrained by biology than Darwin did. Hence *cultural* evolution plays a far more significant role in Marxism than in today's neo-Darwinian theory, according to which human nature is essentially preformed by genetic inheritance, and human culture is severely constrained by the mental architecture of our mind inherited from the Pleistocene. Resisting this kind of biological reductionism, the long-standing tradition of Marxist humanism has sought to develop a more nuanced historical understanding of human nature as a biologically restricted but culturally unbound process: "We have no need to deny that there are some constant attributes constitutive of a human nature," Norman Geras advised leftist theorists back in the 1980s. "It is just not true that this supposition is in itself politically reactionary."[11] To clarify this crucial point, we need to distinguish between "human nature," on the one hand, and the "nature of being human," on the other.[12] In the following, I shall consider "human nature" a sociobiological concept based on neo-Darwinian theory, whereas "the nature of being human" denotes a sociohistorical concept based on the embodied mind and shaped by cultural evolution. So, when I speak of human nature, I intend to emphasize the fixed traits and genetic invariance of the human species. By contrast, when I refer to the nature of being human, I seek to emphasize our species' evolutionary–developmental ability to modify and transcend biological constraints. The reason I have adopted this distinction is not only to avoid conceptual confusions, but also to reflect, on the grammatical–linguistic level, the main difference between neo-Darwinian evolutionary theory, on the one hand, and theories of enactment and the embodied mind, on the other. Simply put, the nominal phrase "human nature" connotes a static being, whereas the gerund "being human" connotes a process of becoming. Using this terminology, I argue that the nature of being human is no more—and no less—determined by human nature than our mind is controlled by the brain and human behavior is explicable in neurophysiological terms alone. Put differently, human nature is a necessary, but insufficient concept for understanding the nature of being human.

In this chapter, I will argue that this distinction, coupled with emergence and system theory and the "new materialisms," allows us to reinvestigate and reconceptualize the relation between Marxist philosophy and Darwinian theory since the late nineteenth century.[13] To begin with, we need to keep in mind that Darwin's own thinking about biological evolution was originally based on metaphorical concepts he borrowed from Thomas Malthus and other sociopolitical discourse at the time. The great Marxist critic Raymond Williams was among the first to highlight the fact that Darwin's theory of biological evolution was originally based on sociopolitical concepts and not the other way around, as the term "social

Darwinism" misleadingly implies.[14] "What Darwin did," Richard Lewontin points out as well, "was take early-nineteenth-century *political* economy and expand it to include all of *natural* economy."[15] The subsequent reapplication of Darwinian terminology back to the sociopolitical realm from whence it had originated was largely due to Darwin's successors rather than Darwin himself. It was Herbert Spencer who popularized Darwin's notion of natural selection and proclaimed "the survival of the fittest" as a natural law in human (social) development. This transfer of biological concepts back into social theory fundamentally altered their erstwhile meaning: "'Fittest,' meaning those best adapted to a given and variable environment," Williams explains, "became 'strongest,' 'most ruthless.'"[16]

Darwin himself, by contrast, emphasized from the very beginning that evolution must be understood in a "large and metaphorical sense including dependence of one being on another," as he stated in 1859.[17] To be sure, in his later work *The Descent of Man* from 1871, Darwin explicitly mentioned the "degeneration" and "injurious" effects resulting from the propagation "of the weak members of civilized societies." Yet he ultimately concluded that "if we were intentionally to neglect the weak and helpless, it could only be for a contingent benefit, with an overwhelming present evil."[18] Throughout his life, Darwin remained reluctant to expand evolutionary theory from biology to human culture and society. Spencer's "conclusions never convinced me," Darwin wrote in his autobiography from 1887, and he reiterated, once again, that "natural selection has been the main, but not the exclusive, means of modification" in human evolution.[19]

These comments may suffice to demonstrate that the history of social Darwinism is inextricably intertwined with the constant migration of scientific concepts back and forth from the social to the biological realm. This conceptual instability of evolutionary theory, in my view, facilitated the rise of twentieth-century euthanasia and eugenics movements across the world and contributed to the racist biopolitics of Nazi Germany and the Holocaust. It also fueled the public controversy about sociobiology in the 1970s and the "science wars" that disrupted the American academy during the 1990s. The next chapter will examine the history of these debates in more detail.

In the present context, I want to focus on another important academic debate regarding the relation of evolutionary theory and social politics that took place in France in the late 1960s. Although ignored by critics since then, the exchange between the Marxist philosopher Louis Althusser and the molecular biologist Jacques Monod helped set the stage for the polemics that erupted a decade later in the United States. Less acrimonious and more substantive by comparison, the French debate centered precisely on the semantic shifts, and ideological valence, of biological concepts if applied to human culture and society. Analyzing this debate will provide key insight into the politico-epistemological forces that keep Marxist philosophy and Darwinian theory apart.

In order to appreciate the arguments on both sides, I shall first introduce the basic tenets of Monod's evolutionary theory based on his 1970 masterpiece *Chance and Necessity*. Monod's book, I argue, tries to bridge the gap between Kant's legacy in premolecular, nineteenth-century biology, on the one hand, and its total eclipse by twentieth-century genetic theory following the modern synthesis in the 1930s and the discovery of DNA in the 1950s, on the other. Being a molecular biologist himself, Monod certainly had no doubts that "the molecular theory of the code" constitutes the quintessential key for understanding life and human development.[20] At the same time, however, and in contrast to many of his Darwinian colleagues, Monod also emphasized that organisms are "*endowed with a purpose or project,*" and he insisted that their purposive behavior (or what he called "teleonomy") "is essential to the very definition of living beings."[21] It is this dual focus on *replication* and *purposiveness,* together with his attempt to provide a post-Darwinian interpretation of Kant's natural philosophy that characterizes Monod's unique contribution to modern evolutionary theory, an accomplishment that neither philosophers nor historians of science have acknowledged or analyzed in any detail so far.

To make my case, the next section develops the central claims of *Chance and Necessity* in the context of both Kantian philosophy and neo-Darwinian orthodoxy. I will try to disentangle the intricate web of scientific and philosophical reasoning that informs Monod's theory and clarify how it relates to our previous discussion of emergence and autopoiesis in chapter 1. Here is my main thesis in a nutshell: much like Kant embraced the notion of the *Bildungstrieb* to split the difference between mechanical and teleological explanations of organized matter, Monod embraced the notion of "invariance" to reconcile the purposive behavior of living beings emphasized in early nineteenth-century, pre-Darwinian biology with the post-Darwinian emphasis of the genetic code in the later twentieth century. The two sections thereafter present a detailed analysis of the Althusser–Monod debate, paying particular attention to its historical setting during the revolutionary spring of 1968 in Paris. The final section of this chapter will showcase the larger significance of this debate in the context of what analytical philosophers call the "norm-circularity" of epistemological paradigms and with regard to Althusser's later writings on "aleatory materialism."

Chance and Necessity, or Kant Revisited

Let me begin at the end. Having rejected the scientific claims of contemporary Marxism, Monod concludes his book with an ethical imperative. The only way we can hope to influence the future of human history, Monod claims, is to develop an "ethics of knowledge" that "is also a humanism." Only "a really *scientific* socialist humanism" can provide the "foundation of social and political institutions" necessary for and dedicated "to the defense, the extension, the enrichment of the transcendent kingdom of ideas, of knowledge, and of creation."[22]

The Kantian roots of Monod's ethical imperative are unmistakable. Although Monod does not make the connection explicit at this point, he frequently refers to Kant throughout his book, most notably when he repudiates Kant's claim about the existence of a priori knowledge, but also when he defends Kant's philosophy against the Marxist charge of idealism.[23] Indeed, not only the end of Monod's book, but its opening pages, too, appear to be written with Kant in mind. Monod starts out with a detailed discussion of the purposive behavior of living organism reminiscent of Kant's distinction between living matter and machines in the *Third Critique*, in particular Kant's admission that it is exceedingly difficult for the human mind to conceive of life in other than teleological terms (though Kant still thought it was wrong to do so, as noted above). Like Kant, Monod, too, confirms "the difficulty of defining the distinction—elusive, for all its obviousness to our intuitions—between 'natural' and 'artificial' objects." In so doing, Monod clearly responds to Kant, if only indirectly: "One of the fundamental characteristics common to all living beings without exception," Monod concludes in the first chapter, is "that of being *objects endowed with a purpose or project*, which at the same time they exhibit in their structure and carry out through their performances."[24] For it would be "absurd," he claims, "to deny that the natural organ, the eye represents the materialization of a purpose—that of picking up images—while this is indisputably also the origin of the camera." Living organisms, no less than human artifacts, pursue a purpose; they are what Monod calls "teleonomic" beings.

This thesis represents a bold move on Monod's part—not because it refers back to Kant, but because it contradicts the by then well-established Darwinian orthodoxy about the non-teleological nature of life that dates back to the great synthesis of Darwin and Mendel in the 1930s. Though Kant lurks in the background, the explicit addressees of Monod's opening gambit are actually his own colleagues, in particular "certain biologists" who still "reject this idea" of teleonomy. The reason why this idea, despite its obviousness, is difficult to accept, Monod argued, is that it gives rise to a fundamental problem—or what Monod called a "*flagrant epistemological contradiction*"—at the heart of modern biology. This contradiction consists in the undeniable "teleonomic character of living organisms," on the one hand, and science's "systematic denial that 'true' knowledge can be got at by interpreting phenomena in terms of final causes."[25] Living beings act purposefully (i.e., they try to mate and stay alive), but life originally emerged and continues to evolve accidentally, without purpose.

Monod, of course, was hardly the first biologist to notice this contradiction. As we saw in the last chapter, Kant, too, pondered the fact that the evolution of life cannot be explained solely in mechanical terms according to the physical laws of nature. In order to account for the ability of living matter to self-organize, Kant was forced instead to stipulate the existence of some inherent drive or immanent purpose at the root of life. He solved the problem, we recall, by declaring the purposiveness of life a regulative principle of reason: he assumed that organic life is teleological and driven by purpose, yet refused to speculate about whether this assumption is actually true and pertains to reality. I also mentioned in the prior chapter that German biologists during the first half of the nineteenth century, though inspired by Kant, abandoned his regulative approach and instead ontologized the *Bildungstrieb* as a constitutive force of life.[26] As the science of biology matured during the second half of the nineteenth century, however, this vitalist notion of materialism became increasingly unpopular due to the rising influence of Darwinian theory and cell theory. Particularly after the discovery of DNA in the 1950s, it became anathema among evolutionary biologists to speak of teleology or to refer to the purposiveness of living organisms. "In order to avoid all the possibility of confusion," Ernst Mayr points out, most biologists today "no longer say that natural selection produces adaptation"—a formulation that implies teleology—"but rather that it results in adaptedness."[27] Needless to say, this strategic reformulation of Darwinian theory failed to solve the problem. All it did was to cover it up.

Monod, in 1970, finally decided to take the bull by the horns. The only way to solve the "flagrant epistemological contradiction" in modern evolutionary theory, he argued in *Chance and Necessity,* was to subordinate the principle of teleonomy to that of reproductive invariance. Although both principles, he emphasized, are necessary for the evolution of life, "invariance necessarily precedes teleonomy." Why? Because the "steady refinement of ever more intensely teleonomic structures are due to perturbations occurring in a structure *which already possesses the property of invariance.*"[28] Living systems, in other words, must *first of all* be stable enough to maintain their specific organizational structure *before* natural selection forces them to develop, *over time,* the kind of teleonomic behavior that characterizes life.

We can appreciate the enormous step forward Monod took in making this claim if we recall that today's proponents of autopoiesis advance a similar critique, namely, that evolutionary theory fails to account for the original emergence of life, which is not based on natural selection, but is based on the chance occurrence of autogenic processes that manage to self-stabilize and resist entropy (see the prior chapter). Monod anticipated this critique almost a decade before Maturana and Varela introduced their seminal concept of autopoiesis in the late 1970s. A crucial difference, however, is that Monod did not consider this insight incompatible with Darwinian theory:

> Of course the theory I am only briefly and dogmatically sketching here is not that of Darwin himself, who could not in his day have had any inkling of the chemical mechanisms of reproductive invariance, nor of the nature of the perturbations these mechanisms undergo. But it is no disparagement of Darwin's genius to note that the selective theory of evolution did not take its full significance, precision, and certainty until less than twenty years ago.[29]

Emergence theory *augments and perfects* evolutionary theory, Monod claimed; it does not challenge or undermine it, as Deacon, Thompson, and many others have argued since. Put differently, Monod accepted the central dogma of neo-Darwinian theory, which defines life in terms of "reproductive invariance" (i.e., DNA), whereas Deacon, Thompson, and others claim that neither invariance nor replication is essential to living structures.[30] There is no way to determine who is right. My point is simply this: the only way for Monod to bridge the gap between emergence and evolutionary theory was to equivocate between the "replicative invariance" of DNA at the molecular level and the "reproductive invariance" of organisms at the species level.[31] This equivocation allowed him, at one and the same time, to claim that "invariance" *precedes* the process of natural selection (i.e., DNA) and that invariance *is the effect* of natural selection (i.e., species).

Another, equally important difference between Monod's theory of reproductive invariance and autopoietic theory is that Monod subordinates teleonomy to invariance as the primary characteristic of living structures: "Ranking teleonomy as a secondary property deriving from invariance—alone seen as primary—the selective theory is the only one so far proposed that is consistent with the postulate of objectivity."[32] Invariance first, teleonomy second—this logical order of events, Monod insists, distinguishes the scientific explanation of life from ideological fantasies like religion, vitalism, animism, et cetera, all of which "assume the reverse hypothesis: to wit, that *invariance is safeguarded, ontogeny guided, and evolution oriented* by an initial teleonomic principle, of which all these phenomena are the purported manifestations."[33]

In autopoietic terms, however, it makes no sense to claim "that invariance necessarily precedes teleonomy." If living structures are *cognitive* autopoietic systems that adapt to their environments—the central tenet of Varela, Thompson, and many other philosophers and cognitive scientists—then teleonomy must be considered a primary, not secondary characteristic of life. If to be alive is to make sense, then even the most primitive forms of life are teleonomic systems that make choices of one kind or another. Even Deacon—who, after all, explicitly rejects the idea that autogenic systems are "cognitive," or "conscious," or "ententional"—emphasizes that such systems are "teleodynamic." That, indeed, is Deacon's main thesis: the autogenic emergence of life is inextricably intertwined with teleodynamic processes at the preorganic, molecular level.[34] By contrast, the ongoing terminological confusion about what to call a bacterium's ability to make choices—is it "immanent purpose" (Kant), or "sentience" (Deacon), or "cognition" (Thompson), or "consciousness" (Margulis and Sagan), or "will" (Varela), or "subjectivity" (Deleuze)?—is entirely secondary and does not change the fact that all life forms are teleonomic by nature. If they were not, they would not be at all.

To repeat: it makes no sense to divorce teleonomy from life. Doing so flies in the face of everything we know, as Monod himself pointed out in the first pages of his book. And yet, in an effort to ward off the specter of vitalism and to reconcile evolutionary theory with the newly discovered molecular theory of the code, Monod felt compelled not just to separate structural invariance from teleonomy, but to link it with reproduction instead (i.e., DNA). From an evolutionary perspective, of course, his suggestion that "the replicative structure of DNA" is the first "self" in the history of life is hardly a contentious claim, but just another way of saying that genes wrote the "book of life." From an autopoietic perspective, however, Monod's notion of invariant reproduction as "the fundamental basis of biology" is a perfect example of the "gene myth" and incompatible even with Monod's own theory at points.[35] What matters instead is that life originates with the being of "individuals" that make choices to stay alive. "The right perspective," Thompson clarifies, "is that of the individual characterization of life."[36] Life makes sense only to the living.

This, essentially, was Monod's view, too, despite his forced attempt to reconcile teleonomy with evolutionary theory. Moreover, *Change and Necessity* challenged orthodox Darwinism in other regards as well, most notably in recognizing the structural independence of cultural evolution vis-à-vis biological evolution. For hundreds of thousands of years, Monod argued, ideational and physical evolution went hand in hand. But with the emergence of language, this coevolution eventually reached a critical threshold that gave rise to what Monod, during his inaugural lecture at the Collège de France in 1967, still called the "noosphere"—that is, the realm of sociocultural evolution that, according to Monod, distinguishes the human species from other animals.[37] Given the unpredictability of human ingenuity and creativity—or what Kant called "the free play of the faculties"—the structural development of the noosphere, Monod concluded, remains scientifically indeterminable. Although *constrained* by our biological nature, the evolution of culture is *not determined* by that nature. Just like the evolution of life is irreducible to physical laws, the evolution of culture is irreducible to biological principles.

Throughout his book, Monod struggled to come to terms with this structural independence of cultural evolution. He needed to tread carefully. For he had already challenged Darwinian orthodoxy with his claim that teleonomy constitutes a fundamental principle in life, going so far as to suggest that "teleonomic performance may indeed be considered decisive in the higher organisms, whose survival and reproduction depend above all upon their behavior."[38] He had also insisted that the first autogenic structure (DNA) must have preceded the process of natural selection, which is to say that life itself had emerged (not evolved) by pure chance. And now he was likely to alienate his Darwinian colleagues even more with the statement that cultural evolution is separate from, and irreducible to, biological evolution. Taken together, these claims could arguably be construed as some version of Lamarckism, the idea that phenotypic experience during a lifetime results in particular traits that are inheritable from one generation to the next. Monod is well aware that these ideas, from an orthodox Darwinian perspective, are close to heresy, yet he implores his audience again and again to remain open-minded and consider the facts only: "An impartial observer, let's say someone from Mars," he stated, "could not fail to be struck by the fact that development of man's specific performance, symbolic language—a unique occurrence in the biosphere—opened the way for *another* evolution, creator of a new kingdom: that of culture, of ideas, of knowledge."[39] This is the second time Monod felt compelled to evoke an extraterrestrial perspective of life on Earth. He had already done so earlier, at the very beginning of his book, to demonstrate the difficulty of distinguishing between machines and organisms. The obvious purpose behind this rhetoric is to emphasize the importance of unbiased observation in science, and to suggest that if his audience finds fault with his logic it is likely due to their own lack of impartiality and the persistence of preconceived notions derived from Darwinian orthodoxy or Marxist ideology. Monod's theory, to repeat, is Kantian through and through. If the Martians he imagines were to utter a word or two, it would undoubtedly be *sapere aude*—the motto Kant used to define the Enlightenment.

And similar to Kant, Monod, too, seems unsure about how far to push his own thoughts on the matter. For there are certain limits he, too, dare not cross. Right after he summoned the Martians to verify the existence of "*another* evolution," he qualifies his previous statement:

> A unique event: modern linguists dwell on the fact that the symbolic language of human beings is of an utterly different order from the various (auditory, tactile, visual, and other) means of communication animals employ. This is no doubt true. But to go on from there to maintain the phenomenon attests to an absolute break in evolutionary continuity—that human language owes nothing whatsoever, even *at the very outset,* to a system of various calls and warnings like those exchanged by apes—this would seem to me a rather difficult step, and in any case an unnecessary hypothesis.[40]

Notice Monod's hesitation at this crucial point: it is "an unnecessary hypothesis," he says, and "rather difficult" to argue that language constitutes "an absolute break in evolutionary continuity"—but he does not say it would be wrong to do so. He simply leaves the issue unresolved. This ambiguity is designed to avoid alienating his Darwinian colleagues even more by questioning the Darwinian principle of natural selection as the fundamental mechanism of life. Instead, Monod offers a compromise: he refers to "a parallel, hand-in-hand development" between physical and ideational evolution, a formulation that anticipates what E. O. Wilson will later call "gene–culture coevolution."[41] Toward the end of his book, Monod once again seeks to reassure his audience about his continued loyalty to Darwin's theory of natural selection: "I do not mean to divide human evolution into two distinct phases," he insists:

> The important point is that during those hundreds of thousands of years, cultural evolution could not help but affect physical evolution; in man more than in any other animal—and owing precisely to its infinitely greater autonomy—it is *behavior* that *orients* selective pressure. And once that behavior ceased to be primarily automatic and became cultural, cultural traits themselves inevitably exerted their pressure upon the evolution of the genome.[42]

Let me emphasize that Monod's argument up to this point is entirely compatible with today's neo-Darwinian perspective on gene–culture coevolution. Like Wilson, Monod stipulates a certain point of historical development—be it the development of language (Monod) or the ability to walk upright (Wilson)[43]—at which cultural evolution fosters a particular kind of behavior—the use of language or the use of hands respectively—"that commits the species irrevocably in the direction of a continuous perfecting of the structures and performances this behavior needs for its support."[44] Once our ancestors started using language, Monod argues, "selection must have favored the development of linguistic ability itself and hence the development of the brain, the organ that serves it." This means that behavior, too, becomes selective in the Darwinian sense of the term, because it goes hand in hand with the evolution of certain physical structures necessary to support it. This process of mutual reinforcement is what Wilson calls "gene–culture coevolution." But then Monod continues:

> This [hand-in-hand development persists], up until the moment when the accelerating pace of cultural evolution was to split completely away from that of the genome. Within the framework of modern societies this split is obviously total. In them, selection has been done away with. Or at least, there is no longer anything "natural" about it in the Darwinian sense.[45]

According to Monod, the split between cultural evolution and natural evolution has grown so vast today as to be "total"—a view directly opposed to neo-Darwinian theory. Although Wilson, too, acknowledges the historical bifurcation of natural and cultural evolution, he insists again and again that this split can never be *total*: "the point must be made," he states emphatically in 1978, that "there is a limit . . . beyond which biological evolution will begin to pull cultural evolution back to itself." Or, as he most famously put it, "the genes hold culture on a leash."[46] This, precisely, is the point of disagreement between Monod and Wilson.

Still, we should note that these two opposing views are not necessarily contradictory; their disagreement results mostly from two different perspectives on life. From a strictly Darwinian point of view, Monod clearly overstated his case. Even though cultural evolution is Lamarckian and based on social learning during lifetime, this process nonetheless requires, and is ontologically based on, the natural selection of living organisms according to Darwinian principles. In order to profit from or contribute to the evolution of culture, a given phenotype must, first of all, be born, which is a biological and not a cultural process. In that narrow sense, indeed, Wilson is right to insist that cultural evolution still depends on biological evolution.

Bare being, however, is just a necessary, but not sufficient, factor to explain human history and culture. Over time, the sociocultural realm, just like the biological realm, generates its own distinct principles of structural development. As cultural evolution progresses ever faster due to technological innovations and increased learning opportunities, human survival and reproduction depend more and more on specific cultural traits that promote social success for the individual, but not necessarily genetic success for the species. Monod's point, once again, was not to deny that selection still takes place in modern society, but to emphasize that "there is no longer anything 'natural' about it in the Darwinian sense." Darwin's theory, after all, operates at the genetic level, but *not* at the level of individual human behavior: "Intelligence, ambition, courage, and imagination are still factors of success in modern societies," Monod readily admits, "but of *personal*, not *genetic* success, the only kind that matters for evolution." In fact, given the increasing independence of cultural evolution in advanced societies, Monod was deeply worried about the increasing gap between human individuals' pursuit of social success, on one side, and the overall fitness of our species, on the other:

> Until not so very long ago, even in relatively "advanced" societies, the weeding out of the physically and also mentally least fit was automatic and

> ruthless. Many of them did not reach the age of puberty. Today, many of these genetic cripples live long enough to reproduce. Thanks to the progress of scientific knowledge and the social ethic, the mechanism which used to protect the species from degeneration (the inevitable result when natural selection is suspended) now functions hardly at all. . . . The only means for "improving" the human species would be to introduce a deliberate and severe selection. Who will want—who will dare to employ it?[47]

I realize that Monod asks this horrific question solely to dismiss it, yet I find the crudeness and insensitivity of his statement, written just twenty-five years after the end of the Holocaust and during a time when eugenics was still legal and common practice in parts of Europe and the United States, extremely troubling even today. Monod was a humanist for sure, and a socialist to boot, yet we shall see in the next chapter that it was but a small step from Monod's scientific belief in "the ethic of knowledge" to E. O. Wilson's claim a few years later that "the planned society" will be "inevitable in the near future" and that ethics, too, will have to be "biologized."[48]

Science and Politics

Chance and Necessity was published in France in 1970. As mentioned above, three years prior, in the fall of 1967, Monod had already presented an early draft of his main ideas during his inaugural lecture at the prestigious Collège de France, which was subsequently published in the French newspaper *Le Monde* on November 30.[49] Around the same time, Louis Althusser, the leading Marxist philosopher in Paris in the 1960s, started his own lecture series on the topic of "Philosophy and Science," held at the École Normale Supérieure from November 1967 to May 1968. In his fourth and final lecture, Althusser seized the opportunity to provide a detailed critique of Monod's inaugural address and the neo-Darwinian theory behind it. Althusser's lectures, however, remained unpublished until 1974, though they were made available in mimeograph form and circulated widely among his students and colleagues before then.[50]

These dates are important, for several reasons. For one, they help clarify the actual order of events, which is skewed by the delayed publication of Althusser's lectures in 1974, even though they were written in 1968. This explains why his critique of Monod only cites the original lecture from 1967, but does not refer to Monod's subsequent monograph from 1970.

Monod's book, on the other hand, not only made reference to Althusser's "severe commentary upon my inaugural lecture before the Collège de France," but also added new and scathing comments about the "silliness" and "epistemological disaster that ensues from the 'scientific' use of dialectical interpretations"—a broadside clearly directed not just against Marxist philosophy in general, but against Althusser's theory in particular, though he did not mention him by name.[51]

Equally important about these dates is the fact that Marxist humanism was a hot topic among French intellectuals at the time. The first complete and apt translation of Marx's *Economic and Philosophic Manuscripts of 1844* into French had just appeared a few years earlier (in 1962), and Althusser had been among the first and most vehement critics of the idealist–humanist tendencies he detected in these early writings by Marx. In fact, Althusser's growing reputation as the leading Marxist philosopher in Paris was founded on his strict opposition to Marxist humanism in favor of what he called "Marx's theoretical anti-humanism" evident in Marx's later works.[52] Based on his nuanced and skillful reading of Marx's *Capital*, Althusser promoted an antihumanist, scientific–materialist version of Marxist philosophy that emphasized, above all, its economic determinism.[53] According to that view, the path of human history, regardless of its ever-increasing complexity, will always be determined "in the last instance" by the economic base (i.e., productive forces and relations of production).

However popular among students and intellectuals at the time, Althusser's radical antihumanist reading of Marx was not unopposed. Most notably, the French Communist Party adopted a formal resolution in 1966 that rejected Althusser's interpretation of Marx along with his critique of the 1844 manuscripts. Instead, the party explicitly confirmed that "there is a Marxist humanism" and that "Marxism is the humanism of our times."[54] Although a member of the party himself, Althusser vehemently opposed this view, which he feared would be disastrous for Marxism both philosophically and politically. In "The Humanist Controversy," an unfinished manuscript composed in 1967 but published posthumously in France in 1993, Althusser provided a detailed critique of "theoretical humanism." His main point was that any and all humanist or anthropological concepts—such as "man" or "the essence or nature of man," as well as terms like "individual," "subject," "genus," "human species," "consciousness," "subjectivity," and even "the notion of labour (as the essence of man)," "the notion of alienation (as the externalization of a Subject)," and "the notion of dialectic (in so far as it implies a teleology)"—are inherently idealist. They serve an "anti-scientific ideological function" in Marx's early writings no less than they did in Kant, Hegel, or Feuerbach. That function, Althusser argued, is to reflect and stabilize the status quo of capitalist society.[55] Hence any philosophy built on or using these concepts will inevitably end up being idealist as well, irrespective of the author's actual intent. The young Marx, in Althusser's view, was a good case in point. Although politically committed to communism, Marx's early texts were hopelessly idealist and still dedicated to "the human individual" and "the essence of man" instead of focusing on the real socioeconomic structure that gives rise to these ideological concepts in the first place. In order to revolutionize society, these idealist concepts, along with the theoretical humanism they stand for, must be expunged from Marxist theory and replaced with the proper materialist concepts Marx developed later in his science of historical materialism (i.e., *Capital*). Hence the term "labour," Althusser claimed, must be replaced by the terms "labour–process" and "labour–power"; and instead of the "essence of man" or the "human subject," Marxists should refer to "the relations of production" as the real and "true 'subjects'" of human history, et cetera.[56]

Althusser's claim that purging Marxist philosophy of idealist concepts was crucial for the fight against capitalism was anything but academic at the time. Instead, his severe critique of Monod was, above all, politically motivated. By exposing the clandestine idealism in Monod's text—a text he considered paradigmatic for "the spontaneous philosophy of the scientists" (SPS) in general—Althusser sought to foster a grand alliance between materialist philosophy and modern science at a time when communist theory and practice was on the rise worldwide. Marxism was being deliberated and waged militarily in Vietnam and Latin America no less than in Poland, Berlin, and Prague. Revolution, indeed, was in the air, and by the time Althusser delivered his final lecture on Monod in the spring of 1968, the revolts had already erupted onto the streets of Paris. The Paris student protests marked one of the most unstable and politically charged periods in the history of the Fifth Republic. Given this revolutionary climate, Althusser felt the need to respond quickly to Monod's highly publicized lecture at the Collège de France, which advocated a decidedly humanist—as opposed to Althusser's decidedly antihumanist—interpretation of materialism and human history.

Hence there was more at stake in this debate than a mere academic dispute between a Marxist philosopher and a molecular biologist about the future of interdisciplinary research. At stake, rather, was the very nature of science itself: its historical origins, its proper methodology, its relation to philosophical ideas, and, above all, its sociopolitical function in capitalist society. At stake, also, was Althusser's contentious claim that dialectical materialism constitutes a genuine science, and that Marx—the "mature" Marx and author of *Capital,* to be sure, not the young, "humanist" Marx still influenced by Hegel and Feuerbach—had uncovered the scientific laws of human history, determined "in the last instance" by economics. And the core issue, for Monod no less than for Althusser, was the right to speak "scientifically"—that is, to speak on behalf of science by usurping its proper voice, the voice of "objectivity"—about the fundamental laws that determine the course of human evolution no less than the survival of France's Fifth Republic.

So what was the gist of Althusser's argument? According to Althusser, science is, by definition, a materialist practice that avows the existence of matter and the objectivity of knowledge. Yet the inherent materialism of science has been consistently undermined "by the religious, spiritualist or idealist–critical 'philosophies of science' manufactured by philosophers or scientists" throughout human history. This history testifies to an eternal struggle between these two conflicting tendencies (or elements) within science: a materialist element genuine to scientific practice (Element 1), and an idealist element (Element 2) foreign to it. The latter interferes with science from the social outside in order to disguise its genuine materialism in ideological terms of human subjectivity and ideational creativity. In his third lecture, Althusser expounds this thesis about the age-old "exploitation of the sciences by philosophy" via historical examples that range from Greek antiquity to the writings of Descartes, Kant, and Husserl.[57] His overall goal is to demonstrate how important it was—and remains today—for modern science to combat Element 2 (idealism) and strengthen Element 1 (materialism).

During the Enlightenment, for example, modern scientists fought back against the dominance of religion and the power of the Church (Element 2) by developing a materialist view of the world based on natural law and Newtonian physics (Element 1). Yet this historical example, according to Althusser, also demonstrates why it is impossible for modern science to shed its idealist tendencies once and for all. For despite their genuinely materialist practice, the rhetoric of eighteenth-century scientists was nonetheless characterized by new idealist concepts like human individualism, historical progress, and bourgeois subjectivity. These idealist concepts, in fact, were no less the product of the eighteenth-century materialist revolt against Church doctrine and authoritarianism as was the new scientific practice they accompanied. Put differently: it was precisely "the juridical, moral and political 'consciousness' of the intellectuals of a rising class" (i.e., a new configuration of element 2) that gave scientists the confidence and courage to challenge religion in the name of a new scientific materialism (i.e., a new configuration of Element 1) in the eighteenth century. The new materialism of modern science was thus made possible by, and remained inextricably interwoven with, the formation of idealist concepts such as "creativity," "individuality," and "subjectivity" that serve to undermine it. This dialectic of materialism and idealism, according to Althusser, characterizes the history of science no less than that of philosophy.

Philosophy, for Althusser, is nothing if not a battleground of ideas *(Kampfplatz der Ideen),* a metaphor he attributed to Kant but used frequently throughout his own writings. It is a place where materialist and idealist tendencies in human thought collide in ever-changing sociopolitical configurations determined by history. The philosophical battle between these two forces amounts to what Althusser called "class warfare in theory." As the historical example of eighteenth-century materialism demonstrates, this war constitutes an eternal struggle without a final resolution or total victory for either side—because neither materialism nor idealism can ever completely eradicate its opponent without, at the same time, eradicating itself. The two are dialectically intertwined forces that require the other's presence to establish its own. In order to intervene in this battle, a genuinely Marxist philosophy must, first and foremost, understand the specific configuration of materialist and idealist tendencies that determine scientific practice at this particular time. Why? Because concepts like "materialism" and "idealism" change their relative meaning and social value throughout human history, and it would be foolish for materialist philosophy to fight idealism without reflecting these changes in theoretical and practical terms of its own. Hence "it is not a question of simply 'applicating' a ready-made philosophy to a determinate SPS," Althusser cautions. Instead, it is "a question of shifting the balance of power within the SPS." The political goal is to challenge the historical dominance of Element 2 (idealist tendencies) over Element 1 (materialist tendencies) so that philosophy may begin to serve, rather than continue to exploit, scientific practice. Yet the "domination of Element 1 by Element 2 cannot be overturned simply through an internal critical confrontation," Althusser insists. "As a general rule, the SPS is incapable of criticizing itself through the play of its internal content alone." In order to realize its own proper materialism, science needs philosophy. But not just any philosophy will do. It must be a genuine materialist philosophy, one willing to modify its own terminology and concepts so as to analyze—and to change—the particular sociohistorical configuration it confronts. Marxist philosophy, too, must adapt to historical change, and it must allow its own concepts to reflect that change if they are to be effective. Only then can materialist philosophy provide the "counterforce" necessary to combat the idealist tendencies that continue to dominate modern science.[58] This is what Althusser sets out to prove with his critique of Monod's inaugural lecture.

For his part, Monod, too, pursued a sociopolitical objective, but one diametrically opposed to that of Althusser. Whereas Althusser's lecture was designed to align Marxism and science in support of historical materialism, Monod sought to separate the two once and for all, because he deemed Marxist philosophy utterly incompatible with objective scientific principles. Monod was well aware of the strong Marxist tradition in the life sciences, a point developed at length by the British biologist J. B. S. Haldane in his book *The Marxist Philosophy and the Sciences* from 1969. Haldane himself made no qualms about why he became a Marxist: "I think that Marxism is true."[59] There can be little doubt that many of Monod's colleagues—not only in Paris, but all over the world, including, for example, Richard Lewontin and Stephen Jay Gould in the United States—shared a similar sentiment at the time.

Monod's fervent opposition to Marxism—and, above all, to Althusser's scientific, antihumanist version of Marxism—forced him to engage in a two-front war: against orthodox Darwinian theory, on one side, and against contemporary Marxist philosophy, on the other. Much like he dealt with the issue of teleonomy, Monod went straight to the heart of the matter regarding Marxism. He emphatically denounced the history of Marxist philosophy—notably Engels and Lenin, exempting Marx himself—for making "dialectical contradiction the 'fundamental law' of all movement, all evolution." Instead of recognizing the historical contingency of both human existence and the knowledge we produce, Marxism, Monod claimed, falsely assumes that nature has "an ascending, constructive, creative intent, a purpose; in short, [Marxism aims] to render nature decipherable and morally meaningful."[60] Marxist theory amounts to a "religion of history": it represents an atavistic remnant of humanity's age-old disposition toward animism, our inherited urge to project purpose and meaning onto the existence of nature and our own being.[61] But there is no meaning or purpose to life, Monod insisted. Life emerged accidentally and entirely by chance, as did the human species.

The contingency of life, Monod argued, explains why the life sciences are irreducible to the physical laws of the material world. These laws are necessary, but insufficient to *predict* the evolution of life, although they can, of course, help *explain* this evolution after the fact. This epistemological gap between the physical and the life sciences is due to the statistical nature of post-Newtonian, scientific knowledge. Monod repeatedly emphasizes that statistical knowledge is oriented toward *classes* of objects and their interrelated development, but does not refer to—and hence cannot predict—the development of any *particular* object or event. And since, according to Monod, "the biosphere does not contain a predictable class of objects or of events but constitutes a particular occurrence," it follows that the evolutionary development of our biosphere, including the appearance of *Homo sapiens sapiens,* is "compatible indeed with first principles, but not *deducible* from those principles and therefore essentially unpredictable."[62]

Much like Monod's theory anticipated some aspects of autopoietic theory, it also formulated key insights of evolutionary epistemology that would be fleshed out later by Karl Popper, Donald Campbell, and Gerhard Vollmer, among others. Chief among them is that higher systems of knowledge are irreducible to the laws and principles of the lower systems on which they are founded. Biological theories that contradict natural laws must be judged false, of course; yet it is impossible to determine in advance what exactly is possible (and what not) within the boundaries of natural law. Scientific knowledge is falsifiable over time, but it cannot be verified apodictically from the bottom up. There is neither "complete understanding" nor "an ultimate explanation" for anything in biology, Vollmer insists.[63] Even the undeniable evolutionary success of the human species does not provide *evidence* of truth, but merely serves as an "indicator of truth."[64] Biology has been successful as a science precisely because of its self-imposed limits of inquiry: it concentrates on living systems and disregards nonliving systems; it tries to keep its distance from both metaphysical and theological questions; and it concentrates on facts, not norms—never mind the fact that science's pursuit of objectivity is itself based on social norms rather than scientific principles, because the essential criterion in biology is fitness, not truth. As Monod put it toward the end of his book: "*The very definition of 'true' knowledge reposes in the final analysis upon an ethical postulate.*"[65]

This scientific fact, Monod concedes, is difficult to accept for rational beings like us, which explains why most people continue to disavow the unpredictability of evolution and the contingency of life. That is bad enough, Monod laments. Marxism, however, not only encourages this dangerous disavowal, but claims to provide "objective" proof for the anthropocentric illusion that human history is determined by scientific principles allegedly discovered by Marx and spelled out in *Capital*. That is absurd (says Monod). For if the evolution of life was, and remains today, entirely unpredictable by science, the same must obviously apply a fortiori to the evolution of human culture and society. Human history, in other words, is irreducible to biological principles in much the same way that these principles are irreducible to the natural laws of physics. Monod is quite adamant about this point: the ideational sphere of human culture, he emphasizes, "stands at a yet greater distance above the biosphere than the latter does above the nonliving universe."[66] There simply are no scientific laws able to predict the long-term future of humanity. The survival of humankind instead hinges on our collective will to take ethical responsibility for the choices we make to protect and improve life on earth.

A Biologistic Theory of History

Let me now discuss the specifics of Althusser's response to Monod in May 1968. Althusser zooms in on two crucial points: first, he vehemently rejects Monod's humanism, the notion that human individuals are free to make ethical choices that determine their own lives and the future of humankind. Second, he argues that Monod, contrary to his claim, does *not* sever cultural evolution from biological principles, but rather provides "a biologistic theory" of human history instead.[67] The two points are linked. Human beings, Althusser argues, are not free subjects, nor are their ideas subject to natural selection. Instead, our beliefs and actions are determined, in the last instance, by the socioeconomic structure we come to inhabit and need to accommodate throughout life.

Before he begins his critique, however, Althusser is careful to emphasize the "unparalleled scientific quality and intellectual honesty" of Monod's text, "its scientific richness, its honesty and its nobility." This lavish praise, I would argue, represents more than just due diligence on Althusser's part vis-à-vis a well-respected colleague and internationally renowned scientist who had just received the Nobel Prize two years earlier. I think it also reflects Althusser's genuine admiration for Monod's intellect and the power of his thought. Most important, however, Althusser makes this statement to clarify right from the start that his critique is directed neither against Monod personally nor against his scientific theory per se (the scholarly merit of which, Althusser readily admits, he is unqualified to judge). Instead, his critique is directed solely against "the philosophical use" of evolutionary theory that comes to light in Monod's text: "In speaking of Monod and citing his declarations, I am not attacking Monod himself but the 'realities' which appear in his own 'consciousness' as so many realities which appear in the 'consciousness' of all scientists, and therefore as so many objective realities independent of the subjective personality of the scientists." Strictly speaking, the object of Althusser's analysis is Monod's text, not Monod's science, let alone the personal views expressed by the "individual" or "subject" called Jacques Monod. Like Monod, Althusser, too, considers objectivity the sine qua non of science, and he repeatedly emphasizes that his own lecture is scientific "in the *most objective* manner possible"—which is to say it is based on the science of historical materialism.[68] In order to live up to this claim of objectivity, Althusser's critique must, first of all, acknowledge and reflect the sociohistorical specificity of the idealist discourse he seeks to expose. Hence he cannot simply attack Monod's spontaneous philosophy (or SPS) from the outside in the name of some abstract, ahistorical notion of Marxist materialism gleaned from his reading of *Capital.* As noted above, a materialist philosophy unable or unwilling to think through the reality of sociohistorical change—like the economic shift from nineteenth-century national-industrial capitalism to twentieth-century global-finance capitalism, or the political shift from traditional disciplinary power to biopower—in new terms and appropriate concepts is neither Marxist nor materialist at all, but ideological through and through. So, to prove his point, Althusser's critique must take place and operate primarily from *within* the specifics of Monod's own text.

An objective–materialist critique of this kind is based on what Althusser, in *Reading Capital,* called a "symptomatic" reading of texts. Akin to deconstruction and informed by Lacanian psychoanalysis, a symptomatic reading uncovers "the unconscious of the text" and divulges the ideological suppositions and hidden tendencies implicit in, yet unacknowledged by, the text itself, let alone its author.[69] Such a reading, Althusser claims, cannot but "pay homage" to Monod's text, because it reveals "the richness" of the idealist tendencies that inform Monod's SPS, his philosophy, and his worldview.[70] But Althusser's reading will also demonstrate that Monod's text knows more than its author, so to speak, because it reveals what remains unsaid and unthought by the "subject" Monod himself. If successful, such a reading would not only undermine the hegemonic status and institutional prowess of positivist science vis-à-vis Marxist philosophy—a critical goal in revolutionary Paris and around the world in 1968. It would also, and more important, demonstrate the secret working of bourgeois ideology in the scientific rhetoric of even the most renowned thinkers at the time—like the Nobel laureate Jacques Monod. Such a reading would constitute "objective" proof for Althusser's thesis that the play of linguistic signification far exceeds the conscious control of human "subject," regardless how much they try to express themselves, their "ideas," or their "individuality."

Like that of any scientist, Monod's text, according to Althusser, is characterized by the two opposing tendencies he already scrutinized in his previous three lectures on SPS: "a *materialist tendency* radiating from the material–objective nucleus and scientific practice itself," on the one hand (Element 1), and "an *idealist tendency* radiating from Monod's ideological position in the face of 'values' implied by the social–political–ideological problems that divide the modern world," on the other (Element 2). The "tension" between these two elements explains "the tendential struggle between idealism and materialism in Monod's philosophy."[71] On the surface, Monod's materialist orientation seems undeniable, since he explicitly rejects the spiritualist and/or vitalist account of "living matter" in favor of the more materialist notion of "living systems" governed by distinct chemico-biological processes (i.e., DNA). This scientific–materialist account of human evolution, however, is undermined by Monod's anthropological–humanist assumption that language created man:

> To say that language created man is to say that it is not *the materiality of social conditions of existence,* but what Monod himself calls "the *immateriality*" of the noosphere, "this realm of ideas and knowledge," which constitutes the real base, and thus the principle of the scientific intelligibility of human history.[72]

This claim is unacceptable for Althusser on two counts: first, because cultural evolution is not, and cannot be, autonomous, since it is determined "in the last instance" by the economic base and the history of class struggle; and second, because language is not—at least not primarily—a medium for man's emancipation from nature, as Monod argues, but an ideological tool for the social interpellation of human individuals into "subjects." I say "at least not primarily" because the two claims—language created man versus language interpellates subjects—are not mutually exclusive, even though Althusser suggests they are. There is much to be said, of course, about the anthropological–humanist claim that language created man, which is central to Western philosophy from Aristotle to Heidegger and has been a focal point of critique in postmodern and posthumanist theory for decades, as we shall see in the final chapter of this book. Yet this anthropological claim, however controversial it may be, is hardly unique to Monod's theory, nor does it contradict Althusser's distinctly political claim that language has also served as a repressive ideological tool throughout human history. Both claims are defensible, in my view, and there is no reason to choose between them. Althusser's claim to the contrary is far from self-evident, and much of his critique depends on how he constructs the alleged contradiction between them.

Inspired by Lacanian psychoanalysis, Althusser considers language constitutive of the symbolic order and the ideological state apparatuses that sustain it. Language is not a tool in the service of, or controlled by, human beings to express their own free will. The reverse is true: human individuals are indoctrinated via language to serve ideology, famously defined by Althusser as "the imaginary relation of . . . individuals to the real relations in which they live."[73] Ideological discourse uses language to hail and interpellate individuals—"Hey, you there!"—as "subjects" in order to maintain the social status quo. Hence, in Althusser's view, the use of language does not create "man." Instead, it creates man's illusion of "subjectivity"—that is, the false belief of human individuals that they are, indeed, free "subjects" in control of their own ideas and behavior. This process of subjective interpellation allows the complex, structured whole of society to produce human "subjects" whose function it will be to reproduce the (real) economic relations that sustain their (imaginary) "subjectivity." This claim—that social systems actively maintain and reproduce themselves—was absolutely crucial to Althusser's scientific understanding of historical materialism. He believed that human "subjects" serve to replicate social structures, because history itself "is a process without a subject or goals." Using an analogy from evolutionary theory, we might put it this way: much like the selfish gene, according to neo-Darwinian biologist Richard Dawkins, uses human individuals as hosts to replicate its ontogenetic code, any given society, according to Althusser, interpellates human individuals as "subjects" in order to reproduce its specific relations of production.

Monod's anthropological view of language as the creator of man, by contrast, puts the human "subject" in charge not only of its own behavior and ideas, but of social history, too. This view, according to Althusser, "represents a conception of history classic since the eighteenth century, since the *Aufklärung*, wherein it is the sciences that are the motor of history, and history is ultimately reducible to the history of knowledges, of the sciences and scientific ideas." This humanist–idealist tendency in Monod's text taints his key concepts, Althusser claims, and it undermines the materialist foundation of science Monod claims to uphold. One example Althusser mentions is "emergence," a key concept in contemporary science. Emergence is commonly defined as the process by which larger entities or patterns with distinct properties emerge from smaller or simpler entities or patterns that do not exhibit these properties in themselves. As discussed in the prior chapter, the emergence of living organisms from inert matter is a perfect example. The term was already familiar to Aristotle, though its modern usage dates back to the end of the nineteenth century. Emergence has become particularly important since the second half of the twentieth century due to the rise of systems theory and nonlinear dynamics.[74]

According to Althusser, however, emergence is an empty concept, a "true *deus ex machina*" in Monod's text: "Each time something new happens—a new idea, a new event—Monod utters the magic word 'emergence.' As a general rule, it might be said that when a concept is used to think everything, it is in danger of not thinking anything at all." If there is a "rational kernel" to this concept at all, Althusser claims, it consists solely in "its theoretical potential . . . full of dialectical resonance." In other words, to the degree that the concept of emergence resonates with that of Marxist dialectics, Althusser lets it stand. Apart from these resonances, however, Monod's use of "emergence" is but a symptom of his SPS. The reason is that Monod's text activates different meanings of the concept without properly distinguishing between them: "Emergence is a double property: reproduction *and* creation. Everything is in the 'and.' For the property of reproduction is one thing, the property of creation another. . . . When Monod causes the term emergence to intervene in his scientific exposé, it is practically always to designate the sudden appearance of new forms; reproduction always remains in the shadows."

At first glance, this seems an odd claim to make given Monod's repeated emphasis of "replicative invariance" (DNA) as the first living structure and the most important mechanism for the evolution of life ever since. But that, precisely, is Althusser's point: Monod conceives of "reproduction" solely in biochemical terms, not in sociohistorical terms. The very idea that social systems actively seek to reproduce themselves is not just absent in Monod's account, but inconceivable in terms of his evolutionary theory. From a neo-Darwinian perspective, it makes no sense to refer to the reproduction of social structures, because such structures are not alive, biologically speaking, and hence cannot possibly be said to "reproduce." Althusser, on the other hand, as well as twentieth-century general systems theory (from Ludwig von Bertalanffy to Niklas Luhmann and Fritjof Capra), makes no qualms about the reproductive capabilities of nonorganic, social systems. The question whether such systems are "alive" is not only irrelevant, in their view, but actually serves to impede or obscure the far more important question about the structural mechanisms by which social systems do, in fact, maintain and reproduce themselves over time. If Darwinian evolutionary theory cannot even recognize, let alone explain these social mechanisms, then, from Althusser's Marxist perspective, it is useless at best and ideological at worst.

This, precisely, is Althusser's key charge against Monod: focused solely on biological mechanisms of replication, Monod completely ignores the obvious power of social institutions and practices—such as the bourgeois family or the education system, the Church or the military—to control and reproduce the social behavior of human beings. Regardless how smart or creative we may have become since the emergence of language, Althusser argues, our personal beliefs and social actions will always be determined, in the last instance, by the economic structure and its symbolic order we are forced to accommodate during our lifetime. Monod, by contrast, "thinks he can account for the content of the social existence of men, including the history of their ideas, as a mere effect of the play of neurobiological mechanisms."[75] And since these mechanisms cannot possibly determine or predict the multifarious ideas that pop up in people's heads, Monod erroneously concludes that our ideas are, indeed, our own, that human subjects have free will, and that the future is ours to shape however we choose. Monod's biologism promotes the idealist illusion of subject-centered cultural processes severed from the real material–structural conditions—meaning the actual social, political, and economic conditions that determine human history and behavior *as opposed to* the molecular conditions encountered by the human genome—that underlie them:

> Indeed, when one has given as the sole material base of the noosphere the biophysiological support of the central nervous system, one has to fill the void of the "noosphere" with the help of the spirit, because there is no other recourse—and certainly no scientific recourse. . . . *A mechanistic use of biological materialism outside biology, in history, has the effect of inverting the materialist tendency into an idealist tendency.*[76]

This is Althusser's key charge: Monod's theory of reproductive invariance illegitimately and unwittingly expands the purview of biological concepts from the study of life to the study of history without realizing that this process "affects their *meaning*, the significance of the use made of them."[77] The result is an idealist–imaginary construct—the spontaneous philosophy of a scientist (SPS)—of the real materialist conditions that determine human behavior.

Althusser's critique thus coincides with the central argument I make throughout this book: that the surreptitious transfer of concepts from one epistemological system to another obscures the specificity of their meaning in different contexts. The explanatory gap between the replication of living beings and the reproduction of social systems cannot be bridged simply by transferring, in a purely mechanistic manner, biological concepts from the molecular to the molar level, from the natural realm of living systems to the social realm of human history. In this sense, I readily agree with Althusser that Monod's theory, presented in the guise of scientific materialism, actually advances "*a biologistic theory of history itself.*"[78] Similar biologistic theories of history and culture have become increasingly prominent over the past few decades, as we shall see in the last two chapters. These theories need to be reined in and assessed critically to determine whether, and how, their claims make any sense at all.

In its worst forms, Althusser suggests, biologistic theories come dangerously close to rehabilitating key aspects of social Darwinism. The most glaring exception to Althusser's otherwise respectful tone throughout his critique occurs precisely in response to Monod's suggestion that the principle of natural selection rules over both organisms and concepts, biosphere and noosphere alike. Here is how Monod put his idea in the book:

> For a biologist it is tempting to draw a parallel between the evolution of ideas and that of the biosphere. For while the abstract kingdom stands at a yet greater distance above the biosphere than the latter does above the nonliving universe, ideas have retained some of the properties of organisms. Like them, they tend to perpetuate their structure and to breed; they too can fuse,

> recombine, segregate their content; indeed they too can evolve, and in this evolution selection must surely play an important role.[79]

Again and again in twentieth-century evolutionary theory, one encounters this ill-conceived notion that concepts evolve like organisms, which is based on a false homology between the historical development of abstract ideas and scientific knowledge, on the one hand, and the biological evolution of living matter, on the other. Like many biologists before and after him, Monod, too, falls prey to this homology when he tries to reassure his Darwinian colleagues that his theory does not amount to "an absolute break in evolutionary continuity." Another example is Richard Dawkins, who gained considerable notoriety in 1976 for his "meme theory." As I will explain in more detail in the next chapter, meme theory suggests that memes (e.g., ideas, beliefs) are cultural replicators much like genes are biological replicators. This is basically the same idea Monod had already introduced a decade earlier—namely, that concepts "evolve" in the Darwinian sense of the term and hence are subject to "natural selection" similar to organisms.

In his response to this idea Althusser did not mince words. Monod, he claimed, provides "an astounding biologistic theory of ideas as endowed with the specific qualities of living species, dedicated to the same function and exposed to the same laws. . . . We fall back with this great avant-garde biologist upon banalities that have existed for more than a century and which Malthus and Social Darwinism charged with ideological energy throughout the nineteenth century."[80] I think Althusser is right on this point. The historical connection he draws between Monod's early articulation of Dawkins's meme theory and the lingering appeal of social Darwinism in the twentieth century is absolutely plausible. We shall see in the next chapter that American biologists, a decade later, will level a similar critique against E. O. Wilson and other neo-Darwinians during the sociobiology debate. The irony, of course, is that Monod, unlike Wilson, had actually pushed the structural independence of cultural evolution as far as he thought possible before admitting, albeit reluctantly, that natural "selection must surely play an important role" in this process as well. Forced to choose between historical materialism or natural selection as the *primum movens* of cultural evolution, Monod ultimately sided with the Darwinians against the Marxists. That decision, however, opened his theory to Althusser's charge of rekindling social Darwinism.

Althusser was not the only Marxist to make this charge. In a public lecture from 1972, Raymond Williams, too, lamented that "the crudest ideas of Social Darwinism, and the crudest analogical interpretations of human relationships in quasi-scientific terms, have been remarkably revived and given extraordinary publicity."[81] Although Williams did not mention Monod by name, it is more than likely that he was familiar with *Chance and Necessity* and had Monod in mind. In another lecture from 1978, Williams once again alerts his audience to the "extraordinary revival of some of the crudest forms of Social Darwinism"—this time, no doubt, directed against E. O. Wilson's *Sociobiology* published in 1975—and insists that these theories "have to be patiently analysed and refuted, point by point."[82] Althusser, of course, had done precisely that ten years earlier in his response to Monod. By 1978, however, Althusser's sober critique had already been forgotten, in large part due to the political events in Paris in 1968, but also because it had been drowned out by the public furor in the United States over Wilson's *Sociobiology*.

The rise of biologism today poses similar ideological and institutional dangers to those analyzed by Marxist critics during the 1970s and 1980s. These dangers, I believe, are real. In response, scholars from all disciplines should resolutely reject the flawed logic of Darwinian orthodoxy and the philosophical cover it provides for theories of Social Darwinism. At the same time, however, humanities' scholars in particular need to think through evolutionary theory in their own terms and develop a non-biologistic view of cultural evolution that does not simply revert to the Marxist chimera of historical materialism. Back in 1968, with revolution in the air and students marching on the streets, Althusser had neither reason nor time to do that. Equipped with objective knowledge about the future course of history, there was no doubt in his mind that the social revolt he witnessed owed nothing to chance but was fully determined, albeit in the last instance only, by the scientific laws of historical materialism: "Practically, then, the emergence/chance couple permits Monod to think as emergences based on *chance* phenomena that are perfectly explicable on the basis of a science of history, whose existence he neither mentions nor even suspects."[83]

It seems fair to say that history, in the end, did not vindicate Althusser's scientific view of it. Immediately following the Parisian uprising, the French parliamentary elections of June 1968 gave Charles de Gaulle's conservative party its greatest victory ever with more than two-thirds of the seats. And two decades later, the entire Eastern Bloc started to disintegrate and finally collapsed entirely in the early 1990s, burying Marxist science and its "perfectly explicable" laws of history underneath the rubble.

On Norm-Circularity and Aleatory Materialism

Before concluding this chapter, let me take stock of the conceptual mutations within and political alliances across different epistemological paradigms we encountered so far. I started out saying that evolutionary theory, from its very beginning in the mid-nineteenth century, has oscillated back and forth between the biological realm of living nature and the social realm of human culture. The resulting ambiguity and semantic instability of Darwinian principles and concepts move to center state in the debate between Althusser and Monod. In their lectures, the two took aim not only at each other, but at their own disciplines as well. Hence Monod opposed his Darwinian colleagues on teleonomy and human cultural evolution, while Althusser broke ranks with his Marxist comrades in rejecting humanism and redefining historical materialism.

This disciplinary unruliness, I think, not only redefined the relation between Darwinian evolutionary theory and Marxist philosophy, but also anticipated the U.S. debate about sociobiology that followed later in the 1970s. We should note in particular that Althusser rejected Monod's theory not because he deemed it *too* determinist or *too* materialist—the key charge Lewontin, Gould, and other Marxists later raised against Wilson's sociobiology and neo-Darwinian theory in 1975—but because it was *not determinist and materialist enough*. Monod's theory was ideological because it lacked any sense of *historical materialism* as distinct from *biological* materialism. Similarly, Monod's unorthodox view about the autonomy of cultural evolution aligned him politically *with* Marxist humanism (both in France and in the United States) yet *against* his Darwinian colleagues, although they, unlike the Marxists, fully agreed with Monod on the scientific issues, most notably that DNA replication is the primary principle of biological evolution. My point is that we get a distorted picture about the complex relation of twentieth-century Darwinian and Marxist theory unless we situate the American debate in the context of the earlier French debate, which neither historians nor philosophers of science have done so far.

Once Monod and Althusser turn their critique against each other, the results are more predictable, largely because their respective positions are rooted in different epistemological paradigms that are mutually exclusive at their core. To wit, Althusser criticized Monod, and scientists in general, for his idealist belief in free will and human subjectivity, while Monod criticized Althusser, and Marxists in general, for his anthropocentric illusion about the predictability of history. These accusations, though rhetorically distinct, are akin in substance. Both are directed against our species'—and our own personal—narcissism in one way or another, and each side projects that narcissism onto the other while, at the same time, pretending to have none of it. This resulting stalemate was due to the head-on collision of two opposing political ideologies (Marxism vs. liberalism) fortified by two distinct disciplinary perspectives on life (philosophy vs. science). Anyone who has ever seen an American presidential debate between a Republican and Democrat knows the drill. Depending on where you stand, the Republicans either defend individual freedom or support capitalist exploitation, while the Democrats either promote social solidarity or destroy small businesses. In saying this, I do not wish to equate science with politics, but to remind us that the difference between the two, though significant, is not as radical as academics like to think.

What the Althusser–Monod debate demonstrated, above all else, is that human nature, and the nature of being human, can be construed just as reductively in terms of Marxist philosophy as in terms of Darwinian theory. Orthodox Marxists' belief in economic determinism, in other words, strikes me as structurally analogous to orthodox Darwinians' belief in genetic determinism. The latter contributes to the senseless debate about the supposed primacy of nature over nurture, while the former fuels the equally futile debate about the alleged supremacy of the economic base vis-à-vis the superstructure. This academic tug of war seems inevitable to the degree that each discipline tries to establish its intellectual and institutional hegemony vis-à-vis the other. In epistemological terms, however, not much is accomplished when one side denounces the other's reductive understanding of human culture and evolution solely to promote its own, equally reductive views instead.

Reductionism, to be sure, is essential to human knowledge. Yet it becomes increasingly counterproductive if pushed too far. Excessive reductionism leads to argumentative gridlock due to the logical incompatibility of a set of first principles and "natural laws" rooted in different epistemological systems. Historians of science and analytical philosophers refer to this dilemma as "norm-circularity." As Willard van Orman Quine explains, "reference is nonsense except relative to a coordinate system," because we "need a background language to regress into" in order to determine the exact meaning or reference of any linguistic or symbolic statement.[84] Quine calls this the "principle of relativity." He elaborates: "What makes sense is to say not what the objects of a theory are, absolutely speaking, but how one theory of objects is interpretable or reinterpretable in another."[85] There are no "facts" outside and beyond the system in which they signify, because there is "no way [the world] is under no description," as the pragmatist philosopher Richard Rorty puts it.[86] Similar arguments in support of norm-circularity have been advanced by Nelson Goodman, Hilary Putnam, Bas van Fraassen, Ian Hacking, and many other scientists and philosophers over the last century.[87]

It is crucial to recognize that one cannot escape from norm-circularity with the help of "objective" scientific experiments or statistical analyses. The latter, too, are imbedded within an overall epistemic structure and are meaningless outside it. In his influential book *The Scientific Image,* the philosopher Bas van Fraassen advocates what he calls "constructive empiricism," a view that explicitly recognizes the degree to which cultural traditions and social customs shape our belief in what constitutes scientific evidence. Van Fraassen concluded that it is "not irrational to commit oneself only to a search for theories that are empirically adequate, ones whose models fit the observable phenomena, while recognizing that what counts as an observable phenomenon is a function of what the epistemic community is (that *observable* is *observable-to-us*)."[88] Another representative of this view is the physicist Andrew Pickering, who challenged the alleged objectivity of scientific experimentation that led to the discovery of quarks in the 1960s: "One can only appeal to the reality of theoretical constructs to legitimate scientific judgment when one has already decided which constructs are real," Pickering claimed.[89] He referred to this process as the retrospective adjudication of validity to scientific theories. Contrary to scientists' self-image as passive recorders of empirical data, Pickering argued that "scientists are genuine agents: doers as well as thinkers, constructors as well as observers," meaning that "theory development is semi-autonomous, and only partially constrained by symbiosis with experiment."[90]

The most radical upshot of this view is that norm-circularity implies not only epistemological relativity, but ontological relativity as well. This belief is one of the pillars of Quine's philosophy. It also constitutes a key point of scholarly disagreement about the work of Thomas S. Kuhn, no doubt one of the best-known proponents of norm-circularity. In his 1962 work *The Structure of Scientific Revolutions,* Kuhn provided a trenchant critique of the methodological underpinnings of modern science, claiming that scientific progress is not based on the gradual unfolding of an ever more precise and comprehensive theory. Rather, it requires a radical change—indeed, a full-fledged revolution—that overturns the basic premises and principles that had hitherto determined what qualified as empirical proof and scientific fact to begin with. Given this fundamental change from one system to another, Kuhn rejected as nonsensical any effort to evaluate the prerevolutionary paradigm according to the newly established truth criteria of the postrevolutionary paradigm. For to argue that one is more or less rational than the other would require reference to an all-comprehensive meta-system of absolute rationality—a system that each party, of course, conceptualizes in terms of their own theory. Most important, some of Kuhn's formulations not only support the epistemological argument about norm-circularity, but also acknowledge its ontological consequences. Kuhn famously concluded at one point that scientific "proponents of competing paradigms practice their trade *in different worlds*."[91] His reference to plural "worlds" has perplexed his readers ever since, because it implies the kind of ontological relativity Kuhn explicitly rejects elsewhere in his work. Most analytical philosophers and scientific realists, therefore, tend to agree that Kuhn did not *really* mean to say what he actually ended up saying, though they keep disagreeing about what precisely he may have intended or should have said instead.[92]

It is tempting to speculate for sure. If, for example, Kuhn had referred to "environments" instead of "worlds," nobody, I think, would have had any reason to complain. Ever since Jakob von Uexküll's influential study on the *Umwelt* (literally "surround-world," but commonly translated as "milieu" or "environment") of animals and human in 1909, the notion that each organism and each species inhabits a specific "niche" or "environment" of its own has been widely accepted in twentieth-century evolutionary theory.[93] Since knowledge obviously forms part of the human environment, it stands to reason that proponents of competing theoretical paradigms practice their trade in different *surroundings* or *environments*. Even orthodox proponents of scientific realism, I venture to predict, would have had little trouble accepting this claim, because it does not challenge their main premise that all environments exist in a single "objective" world as defined by natural science. Kuhn, however, did not speak of different environments. He spoke of different worlds. Figure that.

Althusser did. Surprising at it may seem, Althusser himself developed a similar version of ontological relativism toward the end of his life. A central topic in his later writings—written after his recovery from the mental breakdown he suffered in 1980—was dedicated to what he called "aleatory materialism" or "the materialism of the encounter." In a series of close readings from Epicurus and Spinoza to Heidegger and Derrida, Althusser reconstructs the philosophical history of aleatory materialism, a tradition he juxtaposes to the "materialism of necessity and teleology" adopted by Marx, Engels, and Lenin. Epicurus and Lucretius claim that prior to being, there was no meaning or purpose, no beginning or end; there was only a primordial void of isolated atoms. The material world emerged from the "clinamen"—that is, an unpredictable, but lasting encounter between any two of these atoms. Before this encounter, nothing existed "but abstract elements, lacking all consistence and existence." The encounter creates the foundation of being; it alone dons "the basis for all reality, all necessity, all Meaning and all reason." Althusser connects this ancient philosophy to, among others, Heidegger's famous notion *es gibt*. Commonly (and correctly) translated as "there is," the German original literally means "it gives." The "it," Althusser claims, is Heidegger's name for the initial encounter that becomes the world and determines the nature of being. This, a fortiori, includes *our* own nature, the nature of being human. Human nature, Althusser states in the name of aleatory materialism, results from "*an encounter that has taken form* and become necessary, but against the background of the aleatory of the non-encounter and its forms."[94]

Althusser's philosophy of aleatory materialism is attuned to emergence and contingency; it is a materialism that roots actual being within the virtual world of becoming. His new theory of materialism was clearly influenced by the rise of French deconstruction and event philosophy in the 1970s and 1980s—in particular Deleuze and Derrida, but also Foucault, Badiou, and Rancière. The similarities between their philosophical writings and Althusser's theory of the encounter are all too obvious. Less obvious, but just as important in my view, is the lingering echo of Althusser's debate with Monod in 1968. Back then, Althusser had still defended his view of dialectical materialism as the "science of history" against Monod. A generation later, Althusser had reversed course and embraced Monod's critique as his own. In his interviews from the mid-1980s, for example, Althusser vehemently denounced dialectical materialism as "a philosophical monstrosity" and "theoretical absurdity" invented by Engels, not Marx: "It is not possible to talk about the 'laws' of the dialectic, just as it is not possible to talk about the 'laws' of history. A truly materialist concept of history implies that we abandon the idea that history is ruled and dominated by laws which it is enough to know and respect in order to triumph over anti-history."[95] This radical reversal is stunning, even for a thinker known for his frequent theoretical twists and turns. Althusser, for sure, would not be Althusser—and Marxism not Marxism—if they did not embrace contradictions in theory, including a perennial tendency toward self-contradiction.[96] It would be a serious mistake, however, to dismiss Althusser's final turn toward aleatory materialism as just one among many in his wide-ranging oeuvre. Even less convincing, in my view, is the claim that there was, in effect, no reversal at all, because Althusser's philosophy of the encounter only made explicit what had been implicit, or deliberately hidden, in his earlier work. If instead we take Althusser's reversal seriously, as Warren Montag insists we should, we must evaluate its trajectory not only in philosophical terms, but also situate it within the scientific and sociopolitical context at the time.[97]

From a scientific perspective, Althusser's theory of aleatory materialism clearly reflects the influence of Monod's earlier lecture. All that exists, Monod had argued, is the result of a chance occurrence lost in time. So is the evolution of life and human being. Necessity—the physical laws of material nature and embodied being—emerges only *after* a chance encounter has already taken place and begun to last. Science is objective only because there happen to be some animals—humans—who choose to pursue that kind of knowledge. Having resisted this logic for decades, Althusser finally relented and meticulously realigned this theory of scientific knowledge with the philosophical history of aleatory materialism from Epicurus to Heidegger. Althusser now agrees that prior to the event—the clinamen, the encounter—there are no laws and no necessity in the world. Only "it gives" rise to necessity, meaning that even the laws of historical materialism are contingent and open to radical change pursuant to an event whose emergence and effects are unpredictable by philosophy and science alike. The upshot is that human history has no laws, and that any talk about the "science of history" is absurd at best and authoritarian at worst.

By the time Althusser retraced the forgotten history of aleatory materialism, the political climate also had changed drastically in comparison to the late 1960s, not only in France, but throughout Europe and the world at large. Conservatism ruled in Great Britain (Thatcher), Germany (Kohl), and the United States (Reagan); in France, the socialists (Mitterrand), not the communists, were in power. In this situation, it was important for

Marxism to change its view of history, and equally important that Marxism seize the opportunity to reexamine its own intellectual history in the process. Althusser's philosophy of the encounter fits the bill on both charges. The revolutionary spirit of 1968 had made it politically prudent for him to emphasize Marxism's scientific credentials and the historical inevitability of the global rise of communism. In the mid-1980s, however, as the world economy moved past its latest recession and capitalism became more entrenched than ever in the West, this story no longer seemed credible. There was simply too much empirical evidence to contradict it. This prompted Althusser to develop an alternative version of Marxism, one that removes "rationalist materialism" from its dominant position and emphasizes the history of "aleatory materialism" instead.[98] To rewrite human history in terms of aleatory materialism, Althusser suggests, is to gain insight into the contingency of the world. It gives reason to hope that the world might change again, in a radical manner, due to some other encounter, another event—provided it lasted long enough to create new laws and material necessities of its own.[99] Althusser did not speculate about what this event might be. But his materialist philosophy of the encounter was clearly meant to keep Marxists' spirits high and their hopes alive for the kind of utopian future that rationalist materialism had promised but failed to deliver.

I have reason to believe that some Althusserians might object to my reading of his late philosophy, not least because it turns him into a politically opportunistic thinker. I stand by it. I believe all materialist thinkers are opportunists in that sense. They have to be. Not to think opportunistically is to remain obtuse to the material world and the historical changes that codetermine the way human beings think. Bioaesthetics, too, is opportunistic. It takes advantage of the current focus on STEM research in higher education to advocate for more and better collaboration between the humanities and the sciences. Althusser's oeuvre is exemplary in that regard. The distinctive quality of his thought, and the reason it still remains relevant today, is precisely its political and philosophical commitment to materialism. Opportunist or not, Althusser was a materialist thinker, and a thinker of materialism, throughout his life. His philosophy of the encounter proves that much, and is all the better for it.

3 CULTURAL EVOLUTION

> The essential non-analyzability of atomic stability in mechanical terms presents a close analogy to the impossibility of a physical or chemical explanation of the peculiar functions characteristic of life.
>
> **NIELS BOHR**

Sociobiology

In the summer of 1975, Harvard University Press published Edward O. Wilson's *Sociobiology: The New Synthesis.*[1] The book, and its author, became notorious soon thereafter, and *Sociobiology* remained a subject of intense academic and public debate for years to come. There were various reasons for this notoriety, some related to the scientific and disciplinary questions raised by *Sociobiology,* others having to do with political and moral concerns about its potential influence on social politics in the United States.[2] We already encountered a similar mix of academic and political arguments during the Althusser–Monod debate discussed in the previous chapter. The main difference is that the French debate was extremely limited by comparison and all but eclipsed by the political uprising in Paris in 1968. It has received little, if any, scholarly attention since. The American debate, by contrast, went on for decades, well into the 1990s, and involved both scholars and the public at large. Its huge impact can be explained, at least in part, by the fact that Wilson's voluminous, seven-hundred-page book was far more ambitious and provocative than Monod's inaugural lecture had been. *Sociobiology* received substantial press coverage in the media both before and after it appeared in print.[3] It also proved a huge commercial success, selling more than forty-five thousand copies in just a few years and becoming a standard reference in the field of animal behavior and social organization.

Another difference between the two debates is that Althusser was careful to focus his critique on Monod's text, not the person who wrote it. What mattered to Althusser was that the writer was a scientist, and he explicitly distinguished the author from the human being called Jacques Monod. Althusser's basic claim, we recall, was that language exceeds the control of its users, meaning that linguistic sense exists independently from authorial intention. Like all professional readers, Althusser understood that authors neither know nor determine what their texts say, and that only a symptomatic reading is able to reveal their hidden meaning. He was also aware and mindful of the basic literary distinction between the author of a text and the biographical person behind it. The two are not necessarily identical. The author is an abstract concept that has proved useful for the scientific study of literary texts, whereas the flesh and blood person behind the author is neither a concept nor particularly useful in that regard. Not all writers are publicly known as authors (e.g., consider unpublished or anonymous texts, official proclamations, advertisements, or the use of pseudonyms), and not all authors are, in fact, writers (e.g., consider fictional authors, collective authorship, artificial writing machines, or computer programs).

This crucial distinction was all but lost in the sociobiology debate. Wilson was personally vilified and called a "racist" by students and in the media. He received threats of violence and was physically assaulted by members of the Committee Against Racism during the American Association for the Advancement of Science symposium in February 1978.[4] Though less ad hominem, the scholarly debate was extremely harsh, too. Right after the publication of Wilson's book, several of his colleagues at Harvard, including Stephen Jay Gould and Richard Lewontin, formed the Sociobiology Study Group with the express purpose of resisting *Sociobiology* and limiting its influence within and beyond academia. In collaboration with the Boston chapter of Science of the People, the Sociobiology Study Group published an incendiary review of *Sociobiology* in the *New York Review of Books* on November 13, 1975. They charged that Wilson's book essentially corroborated the "deterministic view of human societies and human action" that "provided an important basis for the enactment of sterilization laws and restrictive immigration laws by the United States between 1910 and 1930 and also for the eugenic policies which led to the establishment of gas chambers in Nazi Germany."[5]

Though Lewontin's first single-authored, scholarly review several months later did not repeat this claim, it was severe nonetheless. Lewontin denounced *Sociobiology* as "a caricature of Darwinism" bent on "vulgar reductionism coupled with the disciplinary imperialism of the more 'basic' sciences that leads biologists to claim that they will reduce the social sciences and the humanities to mere special cases of the evolutionary synthesis." He also charged that Wilson employed a circular logic and relied mostly on analogy and metaphorical language to make his case: "Animal behavior is described by a metaphorical carry-over from human social behavior, and then human institutions are rederived, by a kind of back-etymology, as a special case of the more general animal behavior." Thus, Wilson's theory amounts to "an airtight edifice completely impervious to test." It is unscientific. "The naïve reductionist program of sociobiology," Lewontin concluded, "has long been understood to be a fundamental philosophical error. Meaning cannot be found in the movement of molecules."[6]

Wilson was shocked and outraged by this and similar accusations, in particular the charge that his work, and he personally, sought to reinvigorate the politics of eugenics and social Darwinism, as the Sociobiology Study Group had claimed: "I resent this ugly, irresponsible, and totally false accusation," he shot back against Lewontin and his coauthors a month later in the same venue. In his response, Wilson essentially made two interrelated arguments, which he has repeated numerous times since then. First, he claimed that Lewontin and his colleagues had willfully distorted *Sociobiology,* and he implored the public to investigate "the true context in which the statements were made" in order to discover their "true meaning." The true meaning of what he said, Wilson implies, is precisely what he, the author, intended to say. Since he did not intend to write a political book, let alone revitalize social Darwinism, any reading that claims the contrary is, by default, a misinterpretation. To prove his point, Wilson emphasized that "to date, none of the many other scientists who have reviewed the book in scientific and popular journals has misinterpreted it in this or any other important way."[7]

Wilson's appeal to common sense within the scholarly community backfired, however. As the debate went on, scholars became ever more divided over the "true meaning" of Wilson's text—for good reasons, in my view. For whatever the author's true intentions may have been, a careful reading of *Sociobiology* confirms many of Lewontin's charges. *Sociobiology* does, indeed, promote a "macroscopic view" of human knowledge that lets "the humanities and social sciences shrink to specialized branches of biology." And again, later on: "Scientists and humanists should consider together the possibility that the time has come for ethics to be removed temporarily from the hands of philosophers and biologized." And *Sociobiology* frequently uses metaphorical language to describe the social organization of ants as a "caste system," or to refer to the "warfare" (albeit in quotation marks) that exists both within and between ant species, or to characterize one species *(Formica sanguinea)* as "slave holders," et cetera. *Sociobiology* also claims that "behavior and social structure, like all other biological phenomena, can be studied as 'organs,' extensions of the genes that exist because of their superior adaptive value."[8] Regardless of authorial intention, I think most readers would agree that social institutions and structures are not commonly understood as biological phenomena, because they have little, if anything, in common with living tissue. And to claim that social structures exist because of their superior adaptive value is so dense, so bizarre and unverifiable a notion as to defy comprehension altogether. In his defense, Wilson argued that metaphors are commonly used throughout evolutionary science and are not unique to *Sociobiology*. While this is true, insofar as Darwinian theory has always oscillated between the natural and the social realm (see the prior chapter), it does not render Wilson's metaphorical language and misleading analogies any less problematic; it merely indicates the pervasiveness and severity of the problem throughout the discipline.

Wilson's second line of defense was to turn the tables on his critics and ascribe political motifs to *them*: "What I had said was defensible as science. The attack on it was political, not evidential. The Sociobiology Study Group had no interest in the subject beyond discrediting it," Wilson argued in 2006, repeating a claim he made since the very start of the controversy.[9] To appreciate his argument, we must remember that most of *Sociobiology* (i.e., about 90 percent of the book) was concerned with the social organization of insects, birds, and other nonhuman animals. The ensuing debate, however, focused almost exclusively on the last chapter in his book, which applied the sociobiological perspective developed in the previous twenty-six chapters to the history, culture, and society of human beings. According to Wilson, the omission of 90 percent of the book from discussion indicated that such attacks were motivated by political, not scientific, concerns. The fact that his harshest critics, like Richard Lewontin, Richard Levin, Stephen Jay Gould, and many other members of the Sociobiology Study Group were indeed Marxists provided additional support for this argument. The unspoken motive behind the attack, Wilson claimed, was not just "the challenge [critics] sensed to their own authority as natural scientists devoted to the study of social problems,"[10] but also the collision of evolutionary science and Marxism: "Marxism and other secular ideologies previously rested secure as unchallenged satrapies of scientific materialism; now they were in danger of being displaced by other, less manageable biological explanations."[11]

Decades later, in 1999, Wilson went so far as to argue that his critics' ideological commitment to Marxism effectively rendered science subservient to politics, and he explicitly charged Lewontin with trying to establish "a political test of scientific knowledge."[12] Wilson was undoubtedly right to insist that scientific research, not political doctrines, should determine what counts as scientific knowledge, which is why today's speculative theory of "intelligent design" has no place in science classrooms across the United States. But this does not mean that each and every politically motivated critique of scientific theory is inherently authoritarian or ideological. It just depends on the basic coherence, conceptual clarity, and methodological rigor that informs the critique. Scientific debates, like any other, are inherently political—not in the sense that politics does or should determine the nature of science, as E. O. Wilson feared, but in the sense that scientific concepts inevitably migrate across disciplinary fields and give rise to legitimate ethico-political questions and concerns about the influence of scientific theories on human life and society.[13]

I have already discussed the long-standing history of similar debates about the relation of science and politics in the previous chapter. Suffice it to mention that Wilson's critics seemed altogether unfamiliar with French Marxism and knew nothing about Althusser's critique of science or his detailed response to Monod in 1968.[14] Wilson, too, never once mentions Monod in his book, although he, for sure, was familiar with his work and very likely aware of the vehement critique of Marxism in *Chance and Necessity*. I would even argue that the last chapter of *Sociobiology* was actually inspired by Monod's book.[15] Moreover, since Wilson had served on the search committee that hired Lewontin into Harvard and was well informed about his political commitment to Marxism, I also find it hard to believe Wilson was just "a political naïf" who "knew almost nothing about Marxism as either a political belief or a mode of analysis," as he later claimed.[16] It is more likely that Wilson had genuinely underestimated the methodological and epistemological questions *Sociobiology* unleashed but was ill equipped to answer.

The most problematic aspect of *Sociobiology* is not just that Wilson, like Monod, ends up formulating a biologistic theory of human history, but that he seems to promote some kind of genetic engineering to advance social politics. In his excellent review of the *Sociobiology* debate a decade later, Alexander J. Morin, the director of the Office of Science and Society of the National Science Foundation at the time, makes precisely this point:

> *Sociobiology* is also a manifesto, informed by Wilson's orderly vision of nature and designed to persuade us of its truth, its beauty, and its importance in guiding us to right action. It is the latter aspect of the book that has caused most of the trouble, partly because the reader cannot help responding (if only unconsciously) to the passion that infuses the enterprise as well as to its manifest content.[17]

I think Morin was right: what unleashed the affective response to *Sociobiology* was, above all, Wilson's dire predictions about the future of human society coupled with his passionate rhetoric calling for appropriate action based on genetic data and scientific knowledge. Wilson claimed, for example, that "the planned society" will be "inevitable in the near future," because human social evolution increasingly depends on internalized control mechanisms as opposed to external environmental factors: "When mankind has achieved an ecological steady state, probably by the end of the twenty-first century, the internalization of social evolution will be nearly complete," Wilson stated with confidence. Although he does not specify who will be in charge of the planned society in the future—politicians? technocrats? biologists?—he does clarify that in order to make the right policy decisions, its leaders will have to rely on sociobiology to manage "the genetic accounts" of social behavior.[18]

What does this mean? Take, for example, population drift and the resulting increase in gene flow around the world. The current trend in human migration, Wilson argued in 1975, will eventually lead to a decrease of blood relations within local communities. Since one of the main evolutionary explanations for altruistic behavior among humans is kin selection (blood relatives show altruistic behavior toward their own kin because they share the same genes), Wilson cautioned that the result of fewer blood relatives within local populations "could be an eventual lessening of altruistic behavior" within these communities.[19] Ever since Darwin, evolutionary biologists have conducted similarly abstract thought experiments about the alleged sociobehavioral effects that might result from possible genetic changes within large populations—just recall Jacques

Monod's similar concerns about the increased replication rate of "genetic cripples" in advanced societies (see the prior chapter). What renders these speculations so pointless is their prolonged time frame (at least twenty generations or more) coupled with a singular focus on genetic as opposed to cultural evolution. But even apart from cultural evolution, there exist numerous other, nonadaptive and nonselectional factors on the genetic level that influence the history of human evolution and will upend whatever consequences might result from possible variations of genetic ratios within human communities over the next few hundred years.[20]

Another, equally unwarranted assumption of Wilson's had to do with future efforts to steer people away from aggressive behavior and toward increased levels of cooperation. The problem, Wilson claimed, is that we do not yet know whether human altruism and cooperative behavior toward family and friends is coupled with aggression toward strangers on the genetic level. Hence the effort to decrease aggressive behavior among humans might accidentally result in the correlated yet unintended decrease of cooperative behavior as well: "In this, the ultimate genetic sense, social control would rob man of his humanity," Wilson worried. To be clear, we are in danger of losing our humanity *not* because we shall come to inhabit a brave new world of total control, but precisely the opposite: we might lose our humanity because the future control of social behavior—which is inevitable—is prone to backfire due to our yet incomplete knowledge of the neurobiological processes that determine human behavior. Humanity, in other words, is threatened not by the presence, but by the absence of complete sociobiological control based on genetics: "To maintain the species indefinitely," Wilson concluded, "we are compelled to drive towards total knowledge, right down to the level of the neuron and gene." Ethical considerations that impede or delay this process of acquiring total knowledge will have to be suspended for the time being: "Skinner's dream of a culture predesigned for happiness will surely have to wait for the new neurobiology. A genetically accurate and hence completely fair code of ethics must also wait."[21]

The glib way Wilson dispenses with ethics, his passionate call for dissolving anthropology and other social sciences into the new mega-discipline of sociobiology, and his embrace of "the planned society" of the future characterized by the genetic manipulation of human behavior is as disconcerting today as it was some forty years ago. Wilson's admission, in 1978, that he had completely underestimated "the strength and power of the antigenetic bias that has prevailed as virtual dogma since the fall of Social Darwinism" does little to address the core problem of *Sociobiology.*[22] That problem is not some antigenetic bias or public hysteria about social Darwinism. It is, rather, Wilson's ill-conceived attempt to biologize human history, which ends up promoting genetics as the key for solving sociopolitical problems today and in the future. The core problem is genetic determinism—that is, biologism.[23]

As the controversy raged on, Wilson remained both adamant and conciliatory at the same time: "Contemporary general sociobiology might at best explain a tiny fraction of human social behavior in a novel manner," he admitted in 1978. Yet he still defended the basic principle behind it: "Its full applicability will be settled only by a great deal more imaginative research by both evolutionary biologists and social scientists."[24] That same year, in 1978, Wilson doubled down on his premise and published a second book on the topic, this time focusing entirely *On Human Nature.* Sociobiology, he argued, "is simply the extension of population biology and evolutionary theory to social organization" and "to all aspects of human existence." As strong a proponent of scientific materialism as ever, Wilson emphasized that "life and mind have a physical basis" and "that the visible universe today is everywhere subject to these materialist explanations," meaning that "human nature can be laid open as an object of fully empirical research." The final "question of interest, then," Wilson concluded, "is the extent to which the hereditary qualities of hunter-gatherer existence have influenced the course of subsequent cultural evolution. I believe that the influence has been substantial."[25]

The debate about the proper meaning of evolutionary concepts like adaptation and selection, as well as the applicability of these terms to the study of human cultural development, kept smoldering during the 1980s until it flared up again in the early 1990s when Jerome H. Barkow, Leda Cosmides, and John Tooby published their influential anthology *The Adapted Mind,* a book that signaled the rising prominence of Evolutionary Psychology (EP) across numerous academic disciplines since then. Unlike the earlier debate about *Sociobiology,* the controversy about EP quickly spread beyond the natural and social sciences into the humanities, where it became entangled with the escalating science wars and the Sokal hoax of 1996.[26] The debilitating consequences of this PR debacle for the interdisciplinary study of *Begriffsgeschichte* and collaborative research across the two cultures can hardly be overestimated, in my view. They include not only the ridicule and imminent decline of all forms of "constructivism" in favor of "realism"—an utterly nonsensical juxtaposition—but also the meteoric rise of biologism across all fields in the humanities over the last two decades.

Despite their success—or because of it—sociobiology and EP have also been subjected to veracious criticism from a variety of disciplines. A major argument throughout has been that neo-Darwinists' strong focus on natural selection misconstrues not only the dynamic relationship between genes and their environment, but also that between genes and phenotypes, as well as the complex interaction between natural and cultural evolution.[27] Another shared concern was that sociobiology and EP serve to disavow the progressive communitarian aspects of human evolution in favor of a "selfish" individualism that atomizes society and focuses exclusively on the single unit (e.g., "the" gene, "the" cell, "the" organism) as opposed to the larger social whole these units serve to maintain.[28] Among the harshest critics were Hilary and Steven Rose. In their view, neo-Darwinian theory is a Trojan horse designed to smuggle key aspects of social Darwinism back into the humanities and the study of culture at large. Although purportedly critical of social Darwinism, both sociobiology and EP, Rose and Rose claim, are just "modern and seemingly more sophisticated forms" of it.[29]

In the next two sections, I will first summarize and then critique the basic claims and arguments made by EP in order to familiarize readers with the modular "theory of mind" that informs today's biologistic theories of human culture, art, and aesthetics. We shall see that EP is essentially a neo-Darwinian version of eighteenth-century preformationist theory focused on the brain, and that the critique of EP presents an updated account of eighteenth-century epigenesis—never mind that neither proponents nor critics of EP have noticed this historical connection. My own comments will be informed by the main representatives of EP (e.g., Jerome H. Barkow, David M. Buss, Steven Pinker, Leda Cosmides and John Tooby, and Martin Daly and Margo Wilson) and by four of their most influential critics. They include the biologist Peter J. Richerson, the anthropologist Robert Boyd, the psychologist Michael Tomasello, and the philosopher David J. Buller, whom I have chosen for several reasons: first, because these scholars provide a far more balanced and less polemical account of EP than do Lewontin, Gould, Rose and Rose, and many others (Buller, for example, explicitly rejects many of Gould's arguments as unfounded); second, because these critics hail from different disciplines across the two cultures; and finally, because all four scholars, despite their distinct perspectives, agree on what I consider the essential flaw of EP and neo-Darwinian theory in general. *The Adapted Mind,* I argue, fails to account for the cumulative complexity of cultural evolution, the fact that modern cultures constitute a unique ontogenetic niche within the larger human environment that fundamentally contributes to the ongoing evolvability of the human mind.

Evolutionary Psychology (EP)

Akin to sociobiology, yet focused on the human mind in particular, EP claims that the basic physiological and mental characteristics of our species *Homo sapiens* developed some two hundred thousand years ago, toward the end of the Pleistocene era. This time period provided an "environment of evolutionary adaptation" that left a permanent trace in human behavior and constructed what John Tooby and Leda Cosmides call the innate "architecture of the mind." This mental order is set up by "epigenetic rules" prescribed by genetic inheritance and triggered by environmental stimuli during ontogenesis. Wilson provides a concise summary of this process: "Genes *prescribe* epigenetic rules, which are the neural pathways and regularities in cognitive development by which the individual mind assembles itself. The mind grows from birth to death by absorbing parts of the existing culture available to it, with selections *guided* through epigenetic rules inherited by the individual brain."[30] Anybody familiar with eighteenth-century natural history should recognize these "epigenetic rules" as the neo-Darwinian term for what Kant used to call the "natural dispositions" *(natürliche Anlagen)* inherent in living matter. These dispositions, Kant believed, enable the phenotype to adapt to different environments (see chapter 1). We should also recall that a key problem at the time, and the main reason why Kant grew increasingly dissatisfied with this theory, was the assumption that these dispositions lay *preformed* inside the organism and were not subject to change. Although they provided the natural cause for phenotypic variation and species' ability to adapt, the dispositions themselves were considered non-adaptable, immutable, and preformed.

Today, of course, we know this is not true, and that "epigenetic rules" are themselves the product of adaptations during the evolutionary history of our species. But the curious aspect about EP is that this knowledge makes little difference for how it conceives the relation between the inherited architecture of the human mind and the rapidly changing environment in modern culture. The way we think, according to EP, is essentially preformed by the physiology of our brains and the modularity of our minds, both of which have been fixed for thousands or tens of thousands of years. This mental architecture can change, of course, but only through random genetic mutations selected for greater fitness over dozens of generations, which explains why this ongoing evolutionary process plays virtually no role in recent modern history over the last few hundred years, according to EP. Put differently: although our adapted mind *originally evolved* in response to the Pleistocene environment, EP stresses above all the fact that *our cultural environment* today *is itself a product of that mind* rather than the other way around. "What is genetically specified in adaptations," Tooby and Cosmides argue, "is an economical *kernel* of elements that *guides* the construction and initialization of the machinery through *targeted* interactions with *specific* structures, situations, or stimuli in the world."[31] Note the grammatical subject of the sentence above: the agent is the "kernel of elements" (e.g., the genetic code and epigenetic rules inherited from the Pleistocene) that selectively targets and interacts with "specific structures, situations, or stimuli in the world" outside (e.g., the environment). This means that our mental architecture remains unaffected by the diversity of cultural environments our species has come to inhabit over, say, the last five hundred or even five thousand years, because this time frame is far too short for new adaptations to evolve via natural selection. Even more important is that our inherited mental architecture acts as a sensory–cognitive filter vis-à-vis the world outside in order to keep the human environment relatively stable and consistent over time.

This is a core thesis of EP. For if the human environment were to change drastically between each generation, our inherited mental architecture would not only be useless, but harmful to us. To put it more concisely: unless the human environment remains relatively stable over long periods of time, it would be maladaptive to have developed an adapted mind, because its inherited architecture will prepare the phenotype for an environment that no longer exists—which is to say that natural selection would have never allowed an adapted mind like ours to evolve in the first place. It follows that our adapted mind exists precisely *because* it keeps the human life-world steady and predictable—otherwise, we would not be here at all. And this, in turn, explains why human cultures across time and space must, by necessity, adhere to and essentially reproduce the same kind of environment, or what Tooby and Cosmides call "a single human metaculture":

> There is certainly cultural and individual variability in the exact forms of adult mental organization that emerge through development, but these are all expressions of what might be called a single human metaculture. All humans tend to impose on the world a common encompassing conceptual organization, made possibly by universal mechanisms operating on the recurrent features of human life. . . . In fact, it is only the existence of this common metacultural structure, which includes universal mechanisms specialized to mesh with the social world, that makes the transmission of variable cultural forms possible.[32]

I am not sure Tooby and Cosmides would agree, but EP's "theory of mind," in my view, essentially boils down to a neo-Darwinian version of Kant's subject-object correlationism. Like Kant, EP categorically subjects the (appearance of the) external world to the (innate) architecture of the human mind. The fact that this architecture has evolved via natural selection (as opposed to, say, being a gift from God) is certainly crucial to explain why the adapted mind manages to keep human environments stable, yet utterly irrelevant with regard to its innateness and performance on the phenotypic level.

Still, there are several other differences between EP and Kant's model that are important for bioaesthetics. Unlike Kant, EP considers the human mind not a general problem solver but a domain specific modular system. The adapted mind consists of a large assembly of innate and extremely robust modules virtually unchanged from the time they first evolved some tens of thousands of years ago. Though its mechanism must be activated by external triggers during the critical period of ontogenesis, each module is entirely self-regulatory and its function unaffected by environmental changes. The key point is that each module has evolved to deal with a particular problem only, and it tackles that problem without input from, or cooperation with, any other module present in the mind—much like a single neuron in the brain only responds to a particular kind of stimuli within a given range of parameters inside its receptive field.

A good example is language learning. Although a young child may be exposed to relatively few words or sentences without any grammatical explanations, these few stimuli will usually suffice to activate the child's innate language module such that he or she starts talking.[33] According to EP, there are hundreds or even thousands of these autonomous, domain-specific modules in the human mind dedicated to tasks like language acquisition, face recognition, motor capacity, et cetera. This has come to be known as the massive modularity thesis of EP: "That is, humans have psychological adaptations that contain contentful structure specifically 'about' their mothers, 'about' their children, 'about' the sexual behavior of their mates, 'about' those identified by cues as kin, 'about' how much to care for a sick child, and so on."[34] As this list indicates, some of these modules are more important to EP than others. Indeed, most of its research centers on family kinship, sexual reproduction, and mating preferences to support the universal, neo-Darwinian sense of "human nature" that informs today's biologism.[35]

Another important difference between the Kantian model and EP is that Kant allowed for the human mind to transcend the bounds of reason via aesthetic experience, whereas EP does not. This is not to deny that Tooby and Cosmides explicitly acknowledge both the existence of phenotypic variation across human populations and the influence of different (cultural) environments that trigger these variations: "It is a complete misunderstanding," they insist again and again, "to think that an adaptationist perspective denies or in the least minimizes the role of the environment in human development, psychology, behavior, or social life." At the same time, however, EP does insist that all *human beings* must share an innate "kernel of elements" responsible for us *being human* in the first place:

> However variable cultures and habitats may have been during human evolution, selection would have sifted human social and cultural life (as well as everything else) for obvious or subtle statistical or structural regularities, building psychological adaptations that exploited some subset of these regularities to solve adaptive problems. . . . Our immensely elaborate species-typical physiological and psychological architectures not only constitute regularities in themselves but they impose within and across cultures all kinds of regularities on human life.[36]

It is these "statistical or structural regularities" that provide the basis for EP's universalizing claims about human nature and human metaculture. These terms are shorthand descriptors for quantitative distribution patterns of particular traits and expressions sampled across various human populations and their cultures. In other words, the proper sense of these terms is entirely statistical.

This is important in part because the historical role of statistical analyses for Darwinian evolutionary theory still remains a matter of debate in the field. According to Ernst Mayr, for example, Darwin himself realized that his theory of evolution required an analytical shift from the study of individuals to that of populations, while others insist that Darwin's original method was entirely nonstatistical.[37] Either way, the fact remains that genetics and evolutionary theory today is largely based on statistical data of large populations. Most of these studies employ complex mathematical algorithms run on high-speed computers to create digital environments that simulate stochastic, evolutionary change over long periods of time. The overall goal is to find law-like (i.e., statistical) regularities of large-scale phenotypic variation among arbitrarily defined groups of human populations delineated along various geographical, national, or ethnic lines. Generally speaking, this method works well, and much of today's research on genetic disease or biodiversity would be impossible without the gathering of statistical data. At the same time, however, the discrepancy between the real world (which is infinitely complex) and computer simulations (which are not) means that the resulting data is only probabilistic, not conclusive. Put differently, geneticists' statistical information about the relative chance of trait differences among some arbitrarily defined groups of human population over distinct periods of time does not signify, and cannot possibly predict, the future development or evolutionary potential of any specific phenotype or genotype.

The lack of clarity on this point, I would submit, was a major contributor to the public furor about sociobiology in the 1970s, and it was similarly disruptive for the reception of EP during the 1990s. Here is what Wilson actually said in 1978: "The learning potential *of each species* appears to be *fully programmed* by the structures of its brain, the sequence of release of its hormones, and, ultimately, its genes."[38] Note that Wilson does *not* claim that the learning potential of each human *individual* is fully programmed by its genes. His thesis refers to the genotype, not the phenotype. His theory of human nature, much like that of EP, is rooted in statistical analyses that are of limited use only in trying to understand how learning takes place on the individual level, as Wilson himself pointed out. In the next section, we shall see that the question of how to define an "individual"—can species be conceived as individuals, too?—emerged as a key issue among critics of EP.

Social Learning and Sociogenesis

Let me now engage the critique of EP in more detail. I shall focus in particular on EP's claim that human beings "learn" or develop new cognitive skills solely by adding yet another domain specific module to the preformed architecture of our adapted mind—a process that, according to EP, necessarily requires the long-term stabilization of genetic changes via natural selection. What was other scholars' response to the neo-Darwinian dictum that "genetic natural selection operates in such a way as to keep culture on a leash"?[39]

In their introduction to a collection of essays cowritten over the last thirty years, Robert Boyd and Peter J. Richerson identify the linear–temporal logic of neo-Darwinian theory—genes lead, culture follows—as the main obstacle for thinking through cultural evolution in Darwinian terms. Highly influential today, their work originally encountered stiff resistance, even hostility, from both biologists and social scientists. The main reason for this critique, Boyd and Richerson argue, was the persistent belief in the primacy of "natural" evolution over and against "cultural" evolution. According to this view, natural evolution first fixed the psychological architecture of the human mind during the Pleistocene era, while cultural evolution acts on this architecture and calibrates its default setting to fit today's changed environment:

> The implicit evolutionary scenario seems to be that Pleistocene hominids were just extra-smart chimpanzees, clear social animals in whom *social* learning played a negligible role until the evolution of our brain was more or less complete. *Then* we took up culture, whose evolution is completely controlled by the preexisting evolved mind. *First,* we got human nature by genetic evolution; *then,* culture happened as an evolutionary by-product.[40]

Boyd and Richersen's key argument is that this view fundamentally misconstrues both the recursive relation between nature and culture as well as the increasing structural independence of cultural evolution over time:

> This way of thinking neglects the feedback between the nature of human psychology and the kind of social information that this psychology should be designed to process. . . . The trick is that once social learning becomes important, the nature of the behavior that is available to imitate is itself strongly affected by the psychology of social learning. . . . *Selection acting on culture is an ultimate cause of human behavior just like natural selection acting on genes.*[41]

The last sentence in particular clarifies the specific point of contention between neo-Darwinians and their critics. Both parties acknowledge the existence of feedback loops between natural and cultural evolution—that is, after all, what Wilson's theory of "gene–culture coevolution" is all about.[42] Wilson, for sure, never denied that "cultural evolution is Lamarckian and very fast, whereas biological evolution is Darwinian and usually very slow."[43] He, too, recognized the evolutionary significance of this speed differential in the course of human history, leading to the autocatalysis (or positive feedback loop) of cultural evolution.

Given the amount of misinformation in both popular and scholarly accounts of EP, it bears repeating that the point of contention is *not* about the undeniable fact that cultural evolution outpaces natural evolution, nor is it about the existence of reciprocal feedback between the two or the fact that human phenotypic behavior remains variable and subject to nongenetic forces. Everybody agrees on these points.[44] "Behavior is not explicit in the gene, and mind cannot be treated as a mere replica of behavioral traits," Lumsden and Wilson explicitly stated in 1981, reiterating a point Wilson had already made back in 1975:[45] "Genetic biases," he declared during the height of the sociobiology debate, "can be trespassed, passions averted or redirected, and ethics altered; and the human genius for making contracts can continue to be applied to achieve healthier and freer societies. Yet the mind is not infinitely malleable."[46]

In this quote, too, the last sentence is decisive. It clarifies the core issue of the debate, which concerns the historical and structural limit of the kind of feedback loop that links cultural to natural evolution. Is there a point at which the Darwinian laws of natural evolution must be re-conceptualized such that the temporal logic outlined above—nature first, culture second—becomes irrelevant and ultimately meaningless? All critics of sociobiology and EP answer this question affirmatively. But so did Jacques Monod back in 1970, years before the American debate had even started. Let me quote again the key passage from *Chance and Necessity* discussed in the previous chapter. Gene–culture coevolution persists, Monod argued, "up until the moment when the accelerating pace of cultural evolution was to split completely away from that of the genome. Within the framework of modern societies this split is obviously total. In them, selection has been done away with. Or at least, there is no longer anything 'natural' about it in the Darwinian sense."[47] Once this threshold is crossed, cultural and natural evolution develop along *substantially* different paths. From then on, Boyd and Richerson agree, "human culture participates in *ultimate* causation."[48]

Boyd and Richerson have coined the term "dual inheritance" to describe this phenomenon. In a nutshell, the dual inheritance theory states that human cultures affect the process of natural evolution in an absolute manner, which is to say that "the evolution of culture has led to *fundamental* changes in the way that our species responds to natural selection."[49] A frequently cited example is that regional populations in dairy-producing countries have developed the ability to digest lactose, whereas most other humans have not.[50] Another, less measurable yet far more important instance is the exponential increase in average brain plasticity since the rise of Western rationalism and modern scientific inquiry some five hundred years ago. On the anatomical level, of course, the basic structure of the human brain has not changed at all over the last several thousand years. What has changed in modern times is the brain's ability to support and build more neuronal pathways within this structure. In neurological terms, this means that modern cultural evolution "was accompanied by an increase in the plasticity of the brain's neuronal network, which led in turn to a heightened capacity for learning," as Jean-Pierre Changeux puts it. Since then, he adds, "cultural evolution [has] taken over from biological evolution."[51]

It stands to reason that the average brain plasticity among human populations in advanced societies will continue to increase dramatically over the next few decades. As the pace of cultural evolution accelerates and social pressure on individuals mounts to accommodate more and faster modes of learning, it stimulates our brains to increase their neuronal plasticity to cope with new and unexpected stimuli encountered in ever more complex environments. These epigenetic effects of cultural evolution on the human brain are evolutionarily significant and will, over time, affect genetic change as well. Natural selection, too, will favor brains with a higher potential for plasticity over those with less. It is important to realize that this is a Darwinian, not a Lamarckian, claim: what gets passed on from one generation to the next is *not* a particular experience or specific trait acquired during the lifetime of a single phenotype. Instead, what is passed on and gets selected is a random genetic modification that happened to result in an increased neocortical plasticity of a particular individual, because that individual and its offspring are more likely to be successful in today's advanced society than are others without that modification. This process, however, will take many generations to unfold, which renders genetic changes virtually irrelevant for most humanities research in comparison to the immediate and more drastic *epigenetic effects of cultural evolution for the phenotype.*

To repeat, the "priority" of cultural evolution over natural evolution for bioaesthetics is due to the fact that inheritance is not restricted to genetic transference, but takes place on extra-genetic levels as well: "On the question of the mechanism of transmission," Evelyn Fox Keller clarifies, "measures of heritability are simply silent." She points out that "almost all human traits are transmitted from one generation to the next; but at the same time, let us accept the fact that the mechanisms of transmission are very varied. They may be genetic, epigenetic, cultural, or even linguistic."[52] In this sense, blue eyes are inheritable, but poverty and racism are inheritable, too. Both statements make sense, as long as we keep the semantic plasticity of the term "inheritance" in mind.

This insight distinguishes Boyd and Richerson's model from EP. Boyd and Richerson distinguish between three interrelated levels of human learning that contribute to human diversity on genetic, environmental, and cultural grounds: first, there is *innate learning* based on genetic inheritance; second, there is *individual learning* based on the phenotype's unaided exploration of its particular environment through trial and error; and finally, there is *social learning* based on the observation and imitation of the behavior of others. Although these three levels are not mutually exclusive, but interact with one another in complex ways, it is nonetheless crucial to distinguish social learning from individual learning, because social learning spares the organism the effort of trial and error that can be quite costly and possibly lethal.

Sociobiology and EP, however, do not distinguish between individual and social learning. Neo-Darwinian theory in general fails to account for the specificity of human culture, which constitutes a distinct niche that exists *within* our overall natural environment. Instead, neo-Darwinian theory holds fast to a binary structure: it sutures human culture and natural environment into a single concept (environment) structurally coupled to another concept (organism). According to Boyd and Richerson, however, "this is an error. Because cultural variation is transmitted from individual to individual, it is subject to population dynamic processes analogous to those that effect genetic variation and quite unlike the processes that govern other environmental effects. Combining cultural and environmental effects into a single category conceals these important differences."[53] This lack of conceptual fine-tuning explains why EP represents a far more simplistic and less productive approach to cultural evolution than Boyd and Richerson's dual inheritance theory, particularly from a bioaesthetic perspective. The traditional binaries (of natural vs. cultural evolution, Darwin vs. Lamarck, genesis vs. epigenesis, etc.) are inadequate to conceptualize the many different kinds of (molecular, cellular, organic, social, cultural) environments that interact on the level of human populations and their cultures.

Another serious problem is that EP's rigid notion of "the adapted mind" is too beholden to a top-down, functionalist approach that fails to account for the extraordinary range and speed of ontogenetic cognitive development unrelated to natural selection. The psychologist Michael Tomasello has coined the term "sociogenesis" to describe this phenomenon. Sociogenesis describes the accelerated process of "cumulative cultural evolution" that influences human behavior and cognitive abilities. This process takes two distinct forms: a historical form, meaning that human individuals inherit a broad range of cultural tools (e.g., language, artifacts, behaviors) developed by our ancestors to better cope with our environment; and an ontogenetic form, which consists of real-time, actual collaborations between two or more individuals to solve whatever problems they face together. These collaborative processes, Tomasello emphasizes, "are enabled but in no way determined" by phylogenetic adaptations on the genetic level.[54]

Tomasello's distinction between historical and ontogenetic forms of human collaboration clearly echoes Boyd and Richerson's distinction between individual and social forms of learning. In spite of minor differences between their "dual inheritance theory" and "sociogenesis," the two models fully coincide in emphasizing the distinct contribution human culture makes to the ongoing evolution of our species: "And it is worth noting in this context," writes Tomasello, "that taking these processes seriously enables us to explain not only the universal features of uniquely human cognition . . . but also the particularities of particular cultures, each of which has developed for itself via these historical and ontogenetic processes a variety of culturally unique cognitive skills and products during the last several dozen millennia of human history."[55] This last point is crucial. Tomasello's research moves away from EP's singular focus on human "metaculture" as defined by universally shared practices such as incest taboo, mating rituals, and kinship relations. Instead, he examines the multiplicity of different beliefs and behaviors that characterize individual cultures, each of which, he claims, constitutes a distinct learning environment that contributes to the ongoing development of human cognition.

In his more recent work, Tomasello develops the "shared intentionality hypothesis" to document further the importance of social interaction and cooperation for the evolution of what he claims are distinct cognitive abilities of human beings as opposed to other animals.[56] Along similar lines, an increasing number of cognitive scientists now focus on what they call "second-person consciousness." They aim to develop a new theory of mind based on human social behavior in order to move beyond the traditional binary of first-person, phenomenological perspectives versus third-person, objectivist accounts that has defined the social sciences for more than a century.[57] These recent trends in the field of evolutionary psychology and the concomitant analytical shift from a single human metaculture to the historical variety of different cultures opens up a rich and promising field of research for bioaesthetics to explore the specific contribution of individual artworks and aesthetic practices for human evolution.

It bears repeating that Tomasello, like many other cognitive psychologists today, is highly critical of EP and its massive modularity thesis. His critique matches that of Boyd and Richerson cited above, namely, that sociobiology and EP misconstrue the recursive, nonlinear and nonhierarchical feedback between natural and cultural evolution:

> In general, the basic problem with genetically based modularity approaches—especially when they address uniquely human and socially constituted artifacts and social practices—is that they attempt to skip from the first page of the story, genetics, to the last page of the story, current human cognition, without going through any of the intervening pages. These theorists are thus in many cases leaving out of context formative elements in both historical and ontogenetic time that intervene between the human genotype and phenotype.[58]

The same critique is the core argument in David Buller's seminal study *Adapting Minds: Evolutionary Psychology and the Persistent Quest for Human Nature* from 2005. Buller's title is telling. Whereas Barkow, Cosmides and Tooby refer to *The Adapted Mind,* Buller's *Adapting Minds* replaces the past participle with a gerund and switches from singular to plural. Together, these simple grammatical changes adequately express the main difference between his approach and that of EP. Whereas EP emphasizes, above all, the preformed architecture of the human mind, Buller emphasizes its ongoing adaptability on both the ontogenetic and phylogenic level:

> First, at the population level, human minds are continuously adapting to changing environments over evolutionary time; there continues to be

evolution in human psychological characteristics. And, second, at the individual level, a human mind continuously adapts to changing environments over the course of an individual's lifetime; it is not a static structure preprogrammed with the ways in which it will respond to different experiences it may encounter.[59]

In explicit contrast to EP's massive modularity thesis and its " Swiss Army knife" approach to human cognition, Buller, like Kant before him, conceives of the human mind as a general-purpose problem solver. Which is to say that its central features cannot be reverse engineered with reference to the ancient past, as EP has done consistently. Instead, our adapting minds must be understood in the context of more recent environmental changes as described by sociogenesis and the dual inheritance theory discussed above.

A major consequence of this perspectival shift is that Buller strongly rejects EP's core belief in the universality of human nature. In his view, the very "idea of a universal human nature is deeply antithetical to a truly evolutionary view of our species."[60] Buller's main argument rests on the controversial but recurrent claim in evolutionary biology that species do not constitute natural kinds, as most biologists assume, but should instead be conceptualized as "individuals" in much the same way we think of phenotypes as being individuals. The "species are individuals" thesis was first introduced by Michael Ghiselin, David Hull, and Ernst Mayr in the mid-1970s, and has since then become increasingly popular in evolutionary biology across the fields. In the quote below, Buller explains in detail why the "species are individual" thesis contradicts the ill-conceived notion of human nature. The elegance and conceptual clarity of this passage speaks for itself:

Species, then, are larger-scale individuals than organisms, but they are individuals in the same sense that organisms are. And conspecific organisms are *parts* of the same species, in the same sense in which two cells can be parts of the same body. . . . Thus, when Cosmides and Tooby claim that, "*by virtue* of being members of the human species, all humans are expected to have the same adaptive mechanisms," they are simply wrong. They misunderstand the nature of species, they misunderstand what's involved in two organisms' belonging to the same species, and they fail to understand how the reproductive organization of a species/individual serves to maintain variation among the organism/parts of that species/individual. . . . Shared characteristics are not *definitive* of belonging to the same species, they are

> *incidental* to belonging to the same species. Indeed, since organisms belong to the same species by virtue of being situated within a common genealogical nexus, there need be *no* characteristics that are shared by all the organisms that belong to a species.[61]

Buller's comprehensive and detailed critique of EP has proven extremely influential in the wider field of evolutionary studies. The reason behind this success, no doubt, is his balanced perspective on the matter at hand. Although Buller is sharply critical of EP, he nonetheless remains a strong proponent of evolutionary psychology in general.[62] Many scholars across the disciplines have found Buller's distinction between EP (the neo-Darwinian paradigm) and evolutionary psychology (the academic field at large) extremely useful, and I, too, shall adopt this distinction for the remainder of this book.

Of Memes and Culturgens

It is time to shift focus again. In this chapter so far, we have examined how biological evolution (based on genetic inheritance) differs fundamentally from cultural evolution (based on social learning and sociogenesis). Part of that difference, I noted, is that cultural evolution takes place in many diverse environments that cannot be conceptualized in neo-Darwinian terms. Yet in order to understand how this structural difference expresses itself on the phenotypic level of individual cultural traits, we need to move from the human mind to human artifacts. What, if anything, can evolutionary theory tell us about the genesis and historical development of any particular cultural object? How about specific human behaviors or abstract philosophical ideas, all of which, obviously, are informed by, and contribute to, human culture? And what happens if we chose to short-circuit the complex relation between natural and cultural evolution and transfer neo-Darwinian concepts directly, without any modification or mediation, from the biological to the social realm?

The best way to engage these questions is meme theory. The term "meme" was coined by the neo-Darwinian biologist Richard Dawkins in 1976, who defined it as "a unit of cultural transmission, or a unit of *imitation*." Memes, Dawkins suggested, are cultural replicators analogous to the biological replicators called genes: "Examples of memes are tunes, ideas, catch-phases, clothes fashion, ways of making pots or of building arches."[63] Since memes are replicators in their own right, they give rise to a new and different kind of evolution analogous to, but nonetheless distinct from, genetic evolution.

As Dawkins's idea grew more popular over time, the disgruntlement among evolutionary scientists regarding the nature of memes grew as well. A central problem was that memes, unlike genes, do not necessarily increase adaptive fitness or provide a biological advantage for the human organism—they may, in fact, be maladaptive from the gene's point of view, as Dawkins himself had pointed out early on.[64] Some twenty years later, Susan Blackmore made this point more explicit, claiming that memes might "*appear* to provide advantages even when they do not." The reason why genes and memes can—and frequently do—work at cross-purposes, Blackmore argued, is that memetic replication follows its own evolutionary path independent from that of genetic replication. She readily concedes that memes "could only come into existence when the genes had provided brains that were capable of imitation." But once they exist, memes take "on a life of their own." It follows that "Crick was wrong," as Blackmore summarized the case in 1999: "We are not 'nothing but a pack of neurons'; we are a pack of memes too."[65]

To this day, however, meme scholars continue to disagree on precisely this point, for obvious reasons. For if cultural evolution is as important as Blackmore claims, while, at the same time, it is likely to produce maladaptive results in terms of genetic reproduction, how come the human species continues to prosper instead of going extinct? Trying to solve this problem, most scholars have challenged Blackmore's premise that cultural evolution is, or can be, maladaptive. Hence Kate Distin argues that "most such developments [of cultural change] will be *neutral* with respect to our biological survival," while Robert Aunger goes a step further to claim that cultural evolution is "overwhelmingly adaptive for people."[66] Although Wilson never adopted the term meme in his own studies, he, too, struggled to explain why cultural evolution is not always adaptive from the gene's point of view. In his collaborative work with Charles Lumsden in the early 1980s, Wilson admits that "the correspondence between biological fitness and cultural fitness is far from perfect."[67] This, for sure, is an understatement. The issue is never fully resolved, neither by Wilson nor by anybody else. In their coauthored book *Genes, Mind, and Culture* from 1981, Lumsden and Wilson only mention a few isolated studies (such as the positive influence of cuisine on genetic fitness) to support their overall premise that "the differential contribution [of gene–culture coevolution] leads eventually to the prevalence of the genotypes with the highest fitnesses."[68] They simply left it at that and put their trust in future research. They professed no doubt that "the linkage between biology and culture . . . shows signs of becoming a soluble problem" very soon.[69]

Another serious problem for meme theory is that after forty years of substantial scholarship and popular debate on the subject, it still remains unclear what, exactly, a meme is supposed to be. Are memes cultural artifacts and/or human behaviors in the physical world, or are they abstract ideas inside the human mind? And, if the latter, are they equivalent to mental phenomena—the "qualia" of human consciousness—or should we think of them as nothing but neuronal networks and memory traces in our brains? There are proponents for each of these views in recent meme scholarship.[70]

An early symptom of the problem was that Dawkins himself already began to retreat from his original proposal in the early 1980s. Although he continued to mull over and tweak his original idea for decades, he realized early on that his key concept, the meme, needed to be more clearly defined: "I was insufficiently clear about the distinction between the meme itself, as replicator, on the one hand, and its 'phenotypic effects' or 'meme products' on the other," he stated in 1982. "A meme should be regarded as a unit of information residing in a brain."[71] With this statement, Dawkins abandoned the key component of his original theory, which had defined memes, in strict analogy to genes, as both abstract code and physical carrier in unison. Let us recall that, according to official DNA doctrine, genes are fully autonomous self-replicators. They are supposed to be structural units and functional agents folded into one. Or, as Schrödinger famously put it back in 1947, genes "are law-code and executive power—or, to use another simile, they are architect's plan and builder's craft—in one."[72] We know, of course, that this is not quite true: genes are not, strictly speaking, the sole agents of their own replication, because they require interaction with their immediate environment.[73] Genes, in isolation, do not do anything at all. Their activity is regulated by numerous other agents, both intra- and intercellular. Genes are just one, albeit crucially important, part of a much larger—and extremely complex—biological system that informs the reproduction process on multiple ontological levels, including the molecular, the cellular, the phenotypical, and the social level.

Despite this trenchant critique, the "gene myth" continues to be popular even among those who should know better.[74] After being in place for more than half a century, the DNA doctrine has become far too engrained in today's life sciences to be deposed any time soon.[75] Memetics, by contrast, evidently lacked the historical and institutional support of its older sibling. Unprotected by the central scientific dogma that shielded genetic theory, memetic theory was born a homeless discipline and remained adrift in the institutional void that separates the two cultures. The conceptual problems hidden inside the "gene" became all too obvious once neo-Darwinian theory migrated from the biological to the social realm—that is, once gene theory was extended to meme theory. There was simply too much semantic volatility at the core of meme theory for it to take roots in scientific discourse. This explains why Dawkins, and neo-Darwinians in general, readily conceived of genes as autonomous agents, while, at the same time, he became increasingly reluctant to apply the same definition to memes—even though the concept, in my view, makes just as much (or little) sense in the former as in the latter case. Yet, unlike genetics, memetics was forced to explain its bifurcated nature (conceived as code and carrier, message and medium in unison) to critics from a variety of disciplines. Memetics failed because it was unable to replace the fiction of autonomous self-replication with another story that made sense to the scientific community. The field disintegrated rather quickly. In 2005, the highly publicized *Journal of Memetics* closed down after only eight years of existence online.

Long before then, Dawkins had already grown wary about the unintended subversive power of his own theory. He became concerned that the conceptual volatility of memetics might spill over into gene theory and start corroding its core doctrine of autonomous self-replication. In 1986, a mere decade into the debate, Dawkins cautioned his colleagues that the meme–gene analogy "can be taken too far if we are not careful."[76] Genes, after all, are code and carrier in unison, whereas memes are not. A meme is only "a unit of information residing in a brain," Dawkins clarified.[77] It is a code not a carrier because memes do not build brains, whereas genes (supposedly) do. Many years later, Daniel C. Dennett still made the case that these "new replicators are, roughly, ideas . . . the sort of complex ideas that form themselves into distinct memorable units."[78] So, the reason why memes, unlike genes, are not autonomous self-replicators is because the information they carry lacks a unique physicality of its own. "Nobody really knows what a meme physically is," Dawkins stated bluntly in his foreword to Blackmore's book in 1999.[79]

To solve the problem, Dawkins kept promoting his crucial distinction between "copying the product" and "copying the instructions."[80] He emphasized that memes do not imitate what already exists, but replicate the code that lies underneath. Memes do not copy human behavior, but copy the abstract knowledge (i.e., the mental representation or abstract idea) that informs this behavior. As the British physicist David Deutsch put it, "It is the meaning—the knowledge—that is the replicator."[81] This abstract—or shall we say idealist?—definition of memes, however, raises the question of how we are supposed to identify memetic knowledge independently of the symbols or behavior necessary to manifest it. The problem is that memes, like any information or knowledge, must obviously take shape in one medium or another, and the information they carry—let alone the "meaning" of this information across time and space—necessarily varies between those media. There is no message without a medium, and the medium always affects the message. I realize that neo-Darwinians are likely to reject this premise. But that would be tantamount to Platonism, plain and simple, and it would violate a core principle of scientific materialism.

My key point is that Dawkins's redefinition of memes as "a unit of information residing in a brain" still failed to distinguish between the different kinds of medial carriers that serve to manifest, preserve, and transmit this information. Instead, Dawkins simply linked the existence and replication of memes to just one carrier in particular, the human brain. Blackmore's nuanced yet significant reformulation of Dawkins's original definition made this point explicit: "Memes," she stated in 1999, "are instructions embedded in human brains, *or in artifacts such as books, pictures, bridges or steam engines.*"[82] Although some scholars have since picked up on this point—most notably Robert Aunger in his excellent book *The Electric Meme* from 2002—the vast majority of evolutionary scientists to date remain unconcerned with the need to distinguish the different medial carriers that facilitate the replication and evolution of human culture(s). Boyd and Richerson, too, consider the human brain the most important storage device in human history: "*Culture is information stored in human brains,*" they emphasize.[83] Although they acknowledge that "some cultural information is stored in artifacts," they still believe that "the most important aspects of culture tend to be those stored in our brains," as they put it in 2006.[84]

I believe this view is shortsighted. The replication of cultural knowledge, to be sure, is invariably relayed through the human brain, because accessing or copying data from one storage device to another requires human intelligence and agency. Yet the brain is not the only medium involved in this process. Cultural information can be stored on any number of different memory devices (e.g., paper, electric circuits, chemical emulsions, smooth surfaces of any kind) using a broad array of different notation systems (e.g., numbers, pictures, spoken and written language, symbolic marks of any kind). And each of these media will inevitably shape and affect the (ideal unit of) information it carries differently, namely, according to the particular requirements of its underlying symbolic network and material apparatus. Put in biological terms, this means that any change in a meme's physical environment will inevitably change the meme itself. The constant process of encoding, recoding, and decoding memes necessarily affects (what we may think of as) the abstract knowledge or cultural unit of information the meme is said to carry (whatever *that* might be). The same is true, a fortiori, of human brains. The brain, too, is a specific medium that configures all (meme) ideas according to its unique physiological structure and the complex network of chemical–electric signals traveling between nodes. And since our brains continue to evolve—meaning that each brain is wired differently, and even those of identical twins are not exactly alike—the same meme might trigger different neuronal responses and inform different behavior among different individuals. This explains the high probability that a wide variety of inaccurate copies will emerge during the reproduction and transmission process of memes, which is not the case in genetic replication. Memes are amorphous creatures; they lack the self-identity and replicative power associated with their natural relatives, the genes.

Dawkins's redefinition of memes had important consequences for evolutionary theory and the study of human culture(s). For one, it dematerialized memes as abstract units of knowledge while, at the same time, it essentially confined this knowledge to the human brain. We shall see in the next section that media and cultural theorists have since mounted a vigorous challenge of the last point in particular. Another consequence of Dawkins's meme theory was to render their replication process extremely volatile. Unlike computers, human brains are not hardwired to one another, nor are they identical clones of some original version the way cells are. Human beings (and their brains) communicate indirectly via the use of symbols, not telepathically. So, is there any reason to assume that cultural knowledge (memes) will replicate accurately from one brain to another? I do not think so. David Deutsch makes the same point. According to him, the reproduction of memes is based neither on self-replication nor, strictly speaking, on imitation (as Dawkins and Blackmore claimed), but ultimately depends on human creativity: "The transmission of human-type memes . . . cannot be other than a creative activity on the part of the receiver. Memes, like scientific theories, are not derived from anything. They are created afresh by the recipient."[85] I think that is exactly right, and it reminds me of Kant's notion of aesthetic experience. To identify a meme is to take a leap of faith. It is to presuppose the existence of some idea or abstract knowledge exemplified by the meme yet inconceivable apart from it. If memes exist, they are the product of reflective judgments.

The exemplary nature of memes explains why it is so difficult to define them in abstract, conceptual terms, and why there are so many different meanings attached to the term. To repeat, a meme can be defined as some qualia or abstract idea present in the human mind, or as the neuronal map correlated with this qualia or idea in the human brain. But it also makes sense to think of memes as the material–symbolic record of such ideas stored on a variety of artificial media (like stone plates, or paper, or computers), or to identify a meme as any other kind of human behavior or cultural artifact that seems to represent or explicitly refers to these ideas. Bioaesthetics, I believe, will easily accommodate the unruly nature of memes and their conceptual indeterminacy. Memetics merely confirms what we knew all along, namely, that concepts require context and that meaning emerges from, and remains bound to, the distinct material conditions of particular environments that inform our investigation.

From a scientific perspective, however, the failure of meme theory seems far more troublesome. It clarifies that empirical–quantitative studies of culture make sense only if scientists are able to identify and trace the transmission of a self-identical cultural unit both within and between human cultures. The Darwinian question "How does culture evolve?" necessarily entails the methodological commitment to define a "core element" of (a particular) culture, as Boyd and Richerson rightly point out: "The most interesting outstanding question is the size and timescale of coherent units of culture. Do single cores in an interrelated complex have real histories that reach back five millennia or more?"[86] However defined, these core units of culture need to be "sufficiently protected from diffusion so that phylogenetic analysis is possible," because only "by identifying genuine cultural homologies can one establish the nature of the initial ideational system that was later transformed by historical processes."[87] Once traced, these homologies must be formalized mathematically and their evolution studied with the help of computer-assisted statistical models.

This, precisely, is what E. O. Wilson set out to do in a series of collaborative studies with Charles Lumsden in the 1980s. After 1978's *On Human Nature*—the seminal text that inspired countless versions of biologism across the disciplines since then—Wilson's research increasingly tried to define the essence of culture, much like biologism today feels compelled to define the essence of art. The identification of "the natural units of culture," Wilson argued, was necessary to validate the structural analogy between biological and cultural evolution that informs sociobiology.[88] Though Wilson remained an outsider to the meme debate, he pursued a similar approach to define human cultural evolution in strictly neo-Darwinian terms. He, too, claimed that genetics prescribes the path of cultural evolution not directly, but indirectly via the development of the human brain. Gene–culture coevolution, in other words, is guided by the cognitive schemata and physiological structure of our mental architecture:

> Both the proponents and the critics of human sociobiology now agree that it makes little sense to try to connect genes directly to culture. In contrast, the steps leading from genetic prescription through neural organization and cognition to culture can be realistically visualized so as to render the problem of gene culture coevolution tractable in principle.[89]

Wilson's basic idea was that genes prescribe the emergence of epigenetic rules in the brain, and that these rules organize and control the developmental process of cognition in the phenotype—a process that EP, a few years later, will call the adapted mind. Cognition requires memory, the basic units of which are called nodes. Nodes are functional clusters of neurons in the brain that link together and give rise to mental images and phenomenological experience. And since culture is "ultimately a product of mental activity and hence fully explainable only by means of analysis reaching to the level of brain physiology," it follows that the natural units of culture, in neurological terms, are nodes in the brain.

But that is only half of the story, obviously. For "to define culture properly," Lumsden and Wilson readily concede, "it is also necessary to capture what is intended by social scientists and humanists."[90] It is to their credit that Lumsden and Wilson do not shy away from this challenge, even though we shall see that their definition of culture is ultimately for naught and makes little sense overall. As they put it, culture comprises the whole set of human artifacts, behaviors, and what they call "mentifacts" (i.e., mental phenomena of any kind such as ideas, beliefs, etc.). The unwieldiness of this definition is easy to demonstrate, particularly if we try to keep the various elements in each of the groups apart. A bicycle wheel, for example, constitutes a human artifact of one kind, yet when Duchamps mounts it onto a stool and calls it art, that wheel becomes part of an entirely different cultural object—neither of which, obviously, is identical to the human behavior of riding a (one- or two-wheeled) bicycle. Nor is a physical bicycle the same as the word "bicycle" written on this page or the abstract idea of a bicycle conceived in the human mind. Even Wilson himself acknowledged the difficulty of trying to identify a *particular unit* of culture using empirical methods only. Yet he still believed that its "existence and some of its characteristics can be reasonably inferred" on the basis of his scientific research.[91] Coining their own terminology, Lumsden and Wilson, in 1981, decided to call these basic cultural units "culturgens" rather than "memes," even though the essential concept behind these terms is the same, as the authors themselves pointed out.[92]

In order to study the evolution of culture empirically, Wilson suggested, we need to identify and "search for the basic unit of culture" in the human environment (the culturgen) that correlates to the basic physiological unit of culture (the nodes) in the human brain. Nodes and culturgens are structurally correlated units of cognition located within and outside the organism, respectively, and their continuous interaction enables the flow of information back and forth between mind and world. Although this correlation exceeds a direct, "one-to-one" correspondence, as Lumsden and Wilson repeatedly emphasize, they nonetheless insist that the various "culture types [in human development] arise from the semantic nodes," meaning that "cultural evolution cannot be understood without a knowledge of biological evolution."[93]

So, what exactly are culturgens? In their joint book *Promethean Fire* from 1983, Lumsden and Wilson defined a culturgen as "a relatively homogeneous group of mental constructions or their products. In our classification the manufacture or use of a particular artifact is therefore a culturgen."[94] This seemingly simple, straightforward definition is anything but, and it needs to be unpacked to reveal the conceptual problems hidden within. For Lumsden and Wilson use the term "culturgen" to designate both a "*group* of mental constructions or their products" as well as individual elements *within that group*, such as "the manufacture or use of *a particular* artifact" (my emphasis). The basic unit of culture is thus defined as both a set and/or an element within a set. This conceptual ambiguity, I suggest, is due to the heterogeneous nature of human culture. As noted above, culture includes not just human behavior and artifacts of all kinds (e.g., rituals and ceremonies, laughter, murder, language, art, tools, toolmaking, tool use), but also social institutions (e.g., buildings, archives, laws, rules, official forms and procedures) as well as mental concepts (e.g., symbols, myths, religion, ideas, lies).

To bring order into this chaos, Lumsden and Wilson argued that some traits of human culture (i.e., culturgens) are common to all cultures and thus generic to the whole species, while others are not. To determine which is which, they suggest creating monothetic and polythetic sets of entities so as to quantify the degrees of relationship that exist between these entities. A monothetic set of entities is one wherein all entities share the same diagnostic attributes (e.g., size, shape, color, duration of a process), clearly the most simple case. An example of a polythetic set might be "an array of swords or marriage ceremonies, in which each entity possesses a large number of the attributes of the group" yet "no single attribute is both sufficient and necessary for group membership." Using this method, they defined a culturgen as a monothetic set or any polythetic set of cultural artifacts, behaviors, and mentifacts that share a sufficiently high number of diagnostic attributes (at least 45 percent) among them to be "relatively homogeneous."[95]

The statistical logic behind this model helps to clarify why Lumsden and Wilson had little choice but to define a culturgen in equivocal terms as both a set of elements and as elements within a set. For regardless of which taxonomy or diagnostic attributes they choose to delineate a particular set (i.e., culturgen), its boundaries always remain open to both higher and lower taxa, because its constitutive elements (i.e., behaviors, artifacts, and mentifacts) are not self-same, essential units. Rather, they are sets, too, determined by clusters of other shared attributes, and so on. A culturgen, in short, is a set of a set of a set. It comes into being solely as the result of various statistical procedures used to sort cultural traits into more or less arbitrary groups that are subsequently treated as if they were single units suited for further analysis. Statistical studies of this kind, however, have little to tell us about the individual traits or set of traits they reckon with. Put differently, this model does not allow us to determine whether a particular human behavior, artifact, or mentifact actually belongs to this or that particular culturgen.

A simple thought experiment suffices to make the case. Take, for example, the culturgen "swords," which is an example Lumsden and Wilson themselves chose to exemplify the problem of taxonomy that plagues their theory. "In this example," Lumsden and Wilson coyly ask their readers, “which is the culturgen: all bronze swords, all bronze flange-tanged swords, or all bronze flange-tanged Erbenheim swords?”[96] A good question, indeed—if only they could answer it. But they cannot. Nobody can. For in order to determine whether *a particular sword* forms part of the culturgen “swords,” one would have to assess the number of diagnostic attributes that sword shares with any and all other swords in its group. Obviously, there are infinite ways to go about this, since (an array of) swords comprise(s) an infinite amount of (clusters of) possible attributes—particularly in light of the fact that culturgens, according to Lumsden and Wilson, do not only include material artifacts, but also human behaviors and mental ideas. So how are we to know which empirical object belongs to which culturgen? To wit, is any sharp-edged metal rod to be considered a sword in this context? Is the wooden stick a child wields around in pretend play a sword? How about a picture of a sword with the caption “This is not a sword”—is *that* still a sword, or at least part of the culturgen “an array of swords” as defined by Lumsden and Wilson?

The only way to avoid this taxonomic ambiguity and its infinite regress is to keep reducing culturgens down to an essentialist core that resists any further reduction. Lumsden and Wilson do precisely that. “Thus, while culture might be decomposed in a great many different, academically interesting ways,” they admit, “there is one *natural* decomposition that sustains mental development and cultural evolution.”[97] This *natural* decomposition of culture is determined by the epigenetic rules in the brain as prescribed by our genes: “Just as genes are generative units for the polyribonucleotides and (ultimately) polypeptides, epigenetic rules extract generative units from the information system shared by the members of the society and turn them into knowledge structures and other mental representations, the inward traits of culture.”[98] Culturgens, in other words, are the basic *natural* units of culture because they are intended as such in the mind. To recognize a culturgen is to see one, literally, because whatever artifact, behavior, or mentifact we might encounter in our environment, *once we intend it as a culturgen in our mind,* this cognitive process eo ipso generates the (concept of the) culturgen we seek. Why? Because that, precisely, is what culturgens do: “culturgens generate culture.”[99]

A culturgen, neo-Darwinian theory teaches us, is whatever our cognitive apparatus (guided by epigenetic rules based on genetic inheritance) tends to and conceptualizes *as a culturgen.* And since human culture is ultimately a product of the brain, culturgens are essentially the same as neuronal nodes in the brain: "Culturgens can be mapped into node-link structures in long term memory and in many instances can be treated as identical with them."[100] By means of this circular logic, Lumsden and Wilson manage to reduce culturgens—the complex array of human behaviors and artifacts supposedly out there in the real world—into neuronal clusters inside the human brain. What started out as Lumsden and Wilson's explicit attempt to understand culture not only in scientific terms, but also "as intended by social scientists and humanists" (e.g., as material artifacts like paintings and poems, sewage and sewing, institutions and procedures) ends up being precisely that—namely, culture defined solely as "intended," or a mental construct that resides in people's brains.

Before I move on to the final section of this chapter, I want to clarify again what bioaesthetics can learn from the woefully inadequate, neo-Darwinian attempt to conceptualize the evolution of culture in strict analogy to the evolution of nature without tending to the numerous levels of mediation between the two. The empirico-scientific study of cultural evolution is marred by the following dilemma. It would be naïve to study the transmission of entire cultures as if they were single individuals, because this approach disregards the numerous temporal and spatial breaks and subdivisions within culture(s). It is too abstract and unrefined a view to yield meaningful results. At the same time, however, we cannot just focus on some core units of culture either—first, because such units do not exist in reality and have to be defined arbitrarily; and second, because any algorithm-based, statistical survey of cross-cultural correlations and interactions among such arbitrarily defined units is probabilistic, not conclusive. Some of its results might be based on chance rather than actual homologies between these units, because any cultural unit subjected to in-depth analysis over extended periods of time is bound to break up into ever smaller pieces until, finally, it gets lost from sight altogether. The closer you look at a unit of culture, the greater the distance from which it looks back at you.

Both meme theory and the theory of culturgens demonstrate that there are far too many—mutually interacting—variables at different ontogenetic levels to study human populations and their cultures reductively through single units of culture: "The trouble is that thought and culture are not the sorts of things that can have distinct units at all," Mary Midgley rightly points out. "They do not have a granular structure for the same reason that ocean currents do not have one—namely, because they are not stuffs but patterns."[101] Recent studies in complexity theory strongly support this argument. They indicate that reality consists of dynamic processes, not static units or timeless essences.[102] But even if we disregard the inherently dynamic nature of cultures and focus on single, idealized, core units instead (whatever they might be), the problem remains that cultural evolution, compared to genetic evolution, is less prone to maintain the integrity of such units "due to the hypercombinatorial nature of the cultural transmission system."[103] Memes are not gene-like replicators, because the *transmission* of cultural information from one brain to the next will likely lead to a substantial *transformation* of the original information. Whereas genes compete with each other along an either/or dichotomy (i.e., either a phenotype inherits a mutational gene or not; a particular allele is either dominant or recessive), memes do not: "Cultural replication need not have the same dichotomous character," Boyd and Richerson aptly state. "People can learn and remember more than one variant."[104] This is not to deny that cultural variants compete with one another; it simply means that their mode of competition is different and far more complex. Cultural transmission is biased, because it "depends on what is going on in the brains of imitators," whereas natural selection is "independent of human desires, choices, and preferences."[105]

The continued rise of digital media across the globe since the time these comments were made has clearly acerbated this problem. Today, cultural variations of any (idealized) core unit or meme can be transmitted, received, and modified by anyone, at any time, in any place. The combinatorial possibilities of bits and pieces of cultural information extracted from heterogeneous cultures across time and space—as well as the concomitant increase of possible mistranslations and misunderstandings of this information—have increased exponentially since the World Wide Web was made publically accessible in 1991. Both dissemination rate and transmission speed of cultural information will only continue to accelerate, which renders the empirical study and tracking of individual core units of culture not just difficult, but completely unfeasible, in my view. I certainly appreciate Boyd and Richerson's insight that "there is no central core culture that deserves special attention in phylogenetic analysis. Rather, there are multiple 'cores' and sometimes quite small units whose descent can be usefully traced."[106] But I think their optimism about our ability to trace these units empirically is no longer warranted today, if it ever was. Given the ubiquity of digital tools and the quasi-infinite multiplication of cultural variants all over the world, how would one go about tracing the evolution of a particular meme like, say, a popular song or video clip? What medial and spatiotemporal parameters should be set to limit this study? Why those and not others? It is mind-boggling just trying to conceptualize such a study, let alone execute it.

I have similar reservations about big data analyses in general: "Population thinking is not a universal acid that will dissolve existing social sciences," Boyd and Richerson concede. "But it is a better mousetrap, providing useful new tools that can help solve outstanding problems in the human sciences."[107] While this may be true in the social sciences, I do not think it applies to the study of art or other artifacts. There are good reasons why the humanities provide deep analyses of individual works and their histories, because their focus rests on the particular, not the general. This is not to say that human cultural evolution cannot or should not be studied empirically or reductively at all. But we ought to have no illusions about the speculative nature and epistemological price to pay for the scientific reduction of entire cultures or individual artifacts to some (idealized) core units.

Human culture, in my view, cannot be quantified or measured empirically without losing much of its meaning. I believe it is best studied exemplarily—that is, via a series of ad hoc investigations into specific behaviors, artifacts, and mentifacts that happen(ed) to occur at a particular point in time and space. The subsequent effort to draw meaningful historical connections between these isolated instances can hardly be more speculative or abstract, I would argue, than the positivist approach to analyze cultural evolution with the help of statistical models and mathematical algorithms. If one wants to pursue the latter approach nonetheless, the minimum requirement would be to differentiate more clearly between different units of culture (however defined) and their various modes of replication, transmission, and manifestation across time and space. To do that, sociobiology and EP would have to develop a wide variety of new concepts and distinctions (about different modes of learning, different kinds of memory, different sorts of media involved in this process) in order to move neo-Darwinian theory beyond the crippling reductionism that informs today's biologism.

Technogenesis

I have argued that Dawkins's binary distinction between code and carrier—that is, between memes as abstract units of knowledge located in the human brain versus memes as the material–physical products of that knowledge located in the real world—is woefully insufficient to study human cultures and their history. The same is true of Wilson's bizarre attempt first to distinguish, and then to collapse the distinction, between nodes and culturgens. In this final section, I want to show how deconstruction and media theory approach this problem by shifting focus from the brain to other media and sites of cultural evolution.

In a series of books over the past two decades, the French philosopher and media theorist Bernard Stiegler has challenged the widely shared assumption among evolutionary scientists that the human brain constitutes the primordial and most important medium for storing cultural information. According to Stiegler, it "is the *inorganic organization of memory* that constitutes the essential element, the first coup, engendering all the others and being transformed in transforming all the others in its wake. In this complex, the brain has in fact only a secondary role, in no case a preponderant one."[108] The "*inorganic organization of memory*" comprises the myriad of artificial memory devices human beings use to augment the brain's capacity for storing information. The human brain itself, Stiegler argues, has been shaped by the constant process of uploading and downloading memory from and to external memory devices. The brain, in other words, evolves not only through natural selection at the genetic level, but also on the epigenetic level by reinteriorizing the cognitive power gained though the exteriorization of memory. It follows that humans are essentially prosthetic beings whose history is inextricably intertwined and coevolving with *technics*—understood as the use of our body and hands, the development of language or any other material support system that enables us to make distinctions and improve our capacity for learning.[109]

Similar to Boyd and Richerson's tripartite model for human learning, Stiegler develops a tripartite model for the study of human memory. Let us recall that Boyd and Richerson distinguish between innate learning (based on genetics), individual learning (based on trial and error), and social learning (based on the observation and imitation of others). The distinction between individual and social learning is crucial, they argued, in order to account for the specificity of human culture as a distinct niche in our environment. Even though Stiegler does not refer to their model, he makes a similar distinction: he differentiates between "germinal memory" (based on genetic inheritance), "somatic memory" (based on epigenetic effects due to the phenotype's interaction with its environment), and "epiphylogenetic memory" (based on technics and modern media):

> Consequently, the hypothesis can be formulated that here, in apparent contradiction of the laws of molecular biology, epigenesis exerts a powerful countereffect on the reproduction of the species, channeling or conditioning an essential part of the drive of selection. In this case, the individual develops out of three memories: genetic memory; memory of the central nervous system (epigenetic); and techno-logical memory (language and technics are here amalgamated in the process of exteriorization).[110]

Stiegler's detailed account of three distinct kinds of memory differs from the standard evolutionary model, which recognizes only the first two, commonly referred to as "prenatal genetic memory" (operating on the genetic–molecular level) and "postnatal epigenetic memory" (operating on the epigenetic–somatic level). Stiegler augments this binary model of memory by adding a third category, which he gains by subdividing "postnatal epigenetic memory" into two distinct kinds: somatic memory and epiphylogenetic memory. The former is organic and linked to the individual (i.e., the brain); the latter is inorganic and based on mnenotechnics (i.e., artificial media):

> Epiphylogenetic memory is technics: inscribed in the non-living body. It is a break with the "law of life" in that, considering the hermetic separation between somatic and germinal, the epigenetic experience of an animal is lost to the species when the animal dies, while in a life proceeding by means other than life, the being's experience, registered in the tool (in the object), becomes transmissible and cumulative: thus arises the possibility of a heritage.[111]

The basic effect of epiphylogenetic memory is to facilitate social learning. It allows the phenotype to profit from the experiences and knowledge of others instead of having to rely exclusively on individual trial and error. Yet, whereas Boyd and Richerson understand social learning as the (more or less) direct transfer of knowledge from one brain to another, Stiegler claims that mnenotechnics acts as a crucial intermediary in this process. He also insists that epiphylogenetic memory intervenes not only on the phenotypic level, but also on the level of the genotype, because the development of the neocortex, the most recent anatomical addition to our brains, became possible only with the help of technics. The human brain itself, according to Stiegler, gained shape through the "re-interiorization of human being's technical exteriority."[112]

I want to emphasize that Stiegler's account of "technogenesis" amounts to far more than just a generic history of modern media, and it has little in common with the kind of idiosyncratic media archeology championed by Siegfried Zielinski and others.[113] Stiegler instead relies on cultural anthropology, and the work of Gilbert Simondon and André Leroi-Gourhan in particular, to explore the "analogy between the theories of technical and biological evolution." He is well aware that for this analogy to make sense from an evolutionary perspective, he must consider "the development of technics 'for itself'" and posit "an actual process of *selection of technical archetypes* recalling in singular fashion the play of chance and necessity in molecular biology."[114] This, precisely, is what Stiegler does: he puts "forward the hypothesis of an absolutely new genetic process of selection," namely:

> artificial selection: the selection of mutations exerted at the cortical level in the context of a relation to the original milieu, mediated by the technical apparatus constituting the system of defenses and predation and informing simultaneously the process of individual adaptation and evolution of the entire species, which does not imply a heredity of acquired characteristics, even if that illusion ensues.[115]

Technogenesis, Stiegler emphasizes again and again, is a Darwinian, not a Lamarckian process. It describes the autonomous evolutionary process of organized yet inorganic matter that cannot be conceptualized in either biological or cultural terms alone. One might say that Stiegler takes Wilson's theory of "gene–culture coevolution" at its word. Rather than granting priority to either of the two sides (e.g., "genes lead, culture follows"), his model conceptualizes the coevolution of human and technics, of cortex and tool, in terms of an intermediary: the kind of inorganic, yet organized matter he calls epiphylogenetic memory. Technics, in Stiegler's sense, is neither alive nor dead, but instead constitutes a third state of being beyond this dichotomy:

> The inert, although organized, matter *qua* the technical object itself evolves in its organization: it is therefore no longer merely inert matter, but neither is it living matter. *It is organized inorganic matter that transforms itself in time as living matter transforms itself in its interaction with the milieu. . . .* Matter organized technomorphologically is not passive; *the tendency does not simply derive from an organizing force—the human—it does not belong to a forming intentional that would precede the frequentation of matter, and it does not come under the sway of some willful mastery. The tendency*

> *operates, down through time, by selecting forms in a relation of the human living being to the matter it organizes and by which it organizes itself, where no one of the terms of the relation holds the secret of the other.*[116]

This sounds intriguing, but I confess I am not quite sure what to make of it. On the one hand, Stiegler's central thesis—namely, that technogenesis "is itself nothing other than an artificially accelerated evolution, in which the very nature of evolution is changed"[117]—clearly resonates with similar claims put forward by Boyd and Richerson, Monod, and Changeux. They all agree that cultural evolution has broken free from natural evolution. One may quibble about when exactly this emancipation of culture occurred, whether during the Pleistocene some two hundred thousand years ago or with the start of the Anthropocene just two hundred years ago. Either way, we surely are in the midst of this process now, and Stiegler's theory of technogenesis shines a philosophical light on previously underdeveloped areas in the evolutionary study of culture, in particular the autocatalytic feedback between modern media and the nature of being human.

On the other hand, however, Stiegler's theory seems more radical than previous claims about the autonomy of cultural evolution. If, indeed, technics self-evolves just like nature does—that is, according to Darwinian as opposed to Lamarckian principles, as Stiegler repeatedly points out—does this mean technics gives rise to autopoietic structures that, literally, develop a "life" of their own? As my quotation marks around "life" indicate, this interpretation of technogenesis requires a considerable epistemological shift away from Varela and Maturana's biological understanding of autopoiesis as a constitutive principle of living organisms and toward a more radical, constructivist understanding of autopoiesis in Niklas Luhmann's sense as a self-organizing process operative in social structures as well.[118]

Yet neither autopoiesis nor Luhmann's systems theory play a significant role in Stiegler's argument. Similarly, the meme debate is not mentioned once throughout *Technics and Time*, nor is there any reference to any of the other scholars and theories I have discussed in this chapter. Instead, Stiegler considers the coevolution of human and tool as "something like a technical maieutics" that requires grammatization[119]—a concept he defines not primarily in linguistic terms, but more broadly, in the Greek sense of *grammé*, as the "supplement" or "trace" that informs all systems of signification and life itself:

> The history of the *grammé* is that of electronic files and reading machines as well—a history of technics—which is the invention of the human. As object as well as subject. The technical inventing the human, the human inventing the technical. Technics as inventive as well as invented. This hypothesis destroys the traditional thought of technics, from Plato to Heidegger and beyond. Différance is the history of life in general.[120]

The last sentence makes clear that the beating heart of cultural evolution, in Stiegler's view, is *différance.* Stiegler studied with Derrida, and his account of technogenesis is clearly informed by the logic of deconstruction. Yet Stiegler also tries to sever this logic from its hitherto primary object (i.e., writing) and to extend the purview of *différance* across the entire spectrum of material reality. Indeed, *Technics and Time* is part of a growing trend in the humanities to rethink deconstruction in the context of evolutionary biology. Vicky Kirby, for example, calls on the humanities to "*naturalize* language and its productive energies" for the study of life at large.[121] Whereas most historians of science (like Roof, Keller, and Kay) warn against the metaphorical designation of DNA as "text" or "the code of life," Kirby embraces the textuality of life because, in her view, it is not a metaphor at all:[122]

> Importantly, dispersing literacy (textuality) as the weaving of life itself means not only that 'old' texts remain contemporary and productively alive because never closed off in a past that is simply behind us, but also that the 'texts' of seemingly primeval organisms, or even a supposedly inanimate and lifeless entity such as a photon, become subjects of cognitive and agential entanglement and observational intention.[123]

Kirby's claim about the textual nature of life seems overstated. The idea should be formulated in transcendental, not ontological, terms—namely, as a regulative idea in the Kantian sense: life can be studied *as if it were a text,* regardless of whether this is, indeed, the case. I admit this is a far less exciting proposition than Kirby's, but it might be more sustainable in the long run. A deconstructive perspective on life, as a regulative idea, resonates well with recent empirical research in cognitive science, which likewise emphasizes the material, physiological effects of linguistic metaphors on human cortical development and cognitive abilities. As George Lakoff, Mark Johnson, Jerome Feldman, and others have argued, there exists a crucial link between the forming of linguistic metaphors—based on physical experience—and the growth of neuronal maps in the human brain. "There is now very strong evidence that essentially all of our cultural, abstract, and theoretical concepts derive their meanings by mapping, through *metaphor,* to the embodied experiential concepts," Feldman claims.[124] "Without such metaphors," philosophers Lakoff and Johnson agree, "abstract thought is virtually impossible."[125] If, as these studies suggest, metaphors matter in the sense of grounding human cognition, then *différance* might inform not only the meaning of texts, but "the meaning of the body" as well.[126]

Neo-Darwinian critics and other self-proclaimed realists might dismiss the effort to read nature as text as the archetypical move of deconstruction, reminiscent of Derrida's infamous claim *il n'y a pas de hors-texte*—there is no outside-text.[127] But we ought not to forget that it was scientists' talk about "the book of life" that reintroduced these metaphors into the twentieth-century debate about living matter. I say "reintroduced" because Galileo had already used a similar rhetoric centuries earlier, stating that "the grand book of nature" was written, alas, not in genetic code, but in mathematical symbols. If Stiegler's deconstructive account of human cultural evolution is guilty of illicitly metaphorizing life—and I would much like to hear natural and social scientists speak to this point—he shares this responsibility with a long list of famous scientists using similar metaphors in their description of the world. Bioaesthetics, in any case, would certainly profit, both intellectually and institutionally, if scholars from across the two cultures started debating the textuality of life and its meaning in deconstructive terms.

4 EVOLUTIONARY AESTHETICS

> It has become self-evident that nothing concerning art is self-evident anymore, not its inner life, not its relation to the whole, not even its right to exist.
>
> **THEODOR W. ADORNO**

Art and Nature

Adorno's opening statement in *Aesthetic Theory* still rings true today, almost a half century after its posthumous publication in 1970. It remains a commonplace not only within art history, but across the humanities in general, that traditional efforts to answer the age-old question "What is art?" are a waste of time at best and an ideological power play at worst. The problem is that a definitive theory of art amounts to a contradiction in terms. Art, the philosopher Morris Weitz said, is "an open concept," and there will always be cases "that would call for some sort of *decision* on our part to extend the use of the concept."[1] Jean-Marie Schaeffer agrees: "Art is not an object endowed with an internal essence; like every intentional object it is (becomes) what people make of it—and they make the most diverse things of it."[2] Modern examples range from the historical avant-garde to Warhol's Brillo boxes and Campbell's soup cans all the way to minimalism, conceptualism, environmental art, and the postmodern credo of "anything goes" that would come to dominate the art scene during the 1980s and early 1990s. "Modern art," Boris Groys argues, "overcomes or undermines even the traditional distinction between art and non-art."[3] Hence it is hardly surprising that most philosophers of art over the last fifty years—like Groys, Arthur Danto, George Dickie, Rosalind Krauss, and many others—consider "art" to be infinitely malleable and defined solely by the discursive practices of the various social institutions that, literally, deal with works of art—art galleries and museums, libraries and concert halls, academic departments and the book review section in your local newspaper.[4] Art, like human nature, is what we make of it.

I believe there are good reasons to endorse this institutional perspective on art—reasons that have little to do with "post"-modernist theory, but instead reflect the power of today's globalized financial art market to determine the very nature and value of art.[5] Art is always and necessarily part of a cultural economy, even in so-called primitive or precapitalist societies. There is no art outside of culture, and there is no human culture without some basic form of economic exchange that determines the relative value of the objects around us, including art objects. Unlike other objects, the very existence of an art object depends on somebody declaring it worthy of the predicate "art" in the first place. Art is a meta-object. Its nature, if it has one, is as much discursive as it is material. What made Duchamp's urinal a (scandalous) work of art was his audacity to sign it "R. Mutt" and exhibit it publicly. By contrast, whatever it was the cave painters in Lascaux thought they were doing some sixteen thousand years ago, they were certainly not producing "art" in the modern sense of the term first introduced in the Italian Quattrocento. To argue differently would be to impose Western norms and values onto foreign cultures—a practice many humanists consider politically suspect and reminiscent of the West's colonial history of suppressing other countries, peoples, and cultural traditions.

Over the past several decades, however, an increasing number of scholars from different disciplines have begun to question this institutional understanding of art from a Darwinian perspective. Based on sociobiology and evolutionary psychology, their central premise is that art is part of human nature: "Ideally art is powerful enough to cross cultures; it reads the code of human nature," E. O. Wilson had already declared in 1984.[6] In the words of linguist and cognitive scientist Steven Pinker, "Art is in our nature—in the blood and in the bone, as people used to say; in the brain and in the genes, as we might say today."[7] In her 1992 book *Homo Aestheticus,* the renowned anthropologist Ellen Dissanayake points out that contemporary aesthetic theory applies at most to the last 2,500 years of Western culture, and likely only to modern art over the last two centuries.[8] The nascent field of evolutionary aesthetics, by contrast, regards art as a universal characteristic of the entire human species that developed over hundreds of thousands of years ago during the Pleistocene era, circa

200,000 BCE, when our species *Homo sapiens* acquired its basic physiological and mental characteristics by adapting to the environmental conditions prevalent at the time.[9] Given the slowness of natural selection and human genetic change, this time period provided an "environment of evolutionary adaptation" (or EEA) that left a permanent trace in the "architecture of the mind." This mental architecture, according to evolutionary aesthetics, continues to influence, and may even determine, our basic emotional and cognitive abilities and preferences—including what Dennis Dutton, in a recent book, has called *The Art Instinct.*

Dutton is a strong proponent of orthodox EP and its massive modularity theory of the adapted mind: a robust architectural structure that houses a multitude of domain specific modules to solve the kind of problems our ancestors were likely to encounter in their prehistoric environment. In the previous chapter, I have already discussed the numerous objections raised by evolutionary scholars against the reductionist model of EP, none of which is mentioned in Dutton's account. Instead, he claims to provide empirical evidence for our mental architecture by highlighting people's universal interest in calendar art as opposed to high art. In particular, he cites a 1993 study by the Russian artists Komar and Melamid that asked people all over the world to describe their favorite landscape, which was then painted by the authors to reflect the statistical average of responses they received. In Dutton's view, this study convincingly demonstrated that "people in very different cultures around the world gravitate toward the same general type of pictorial representation: a landscape with trees and open areas, water, human figures, and animals."[10]

According to Dutton, the reason why most people almost everywhere enjoy such pictures more than others is evolutionary, meaning that our aesthetic preferences are universal and not relative, biologically innate and not culturally acquired or socially constructed.[11] This belief prompts Dutton to develop a list of twelve essential characteristics that, in his view, define the nature of art—even though he readily concedes that not each and every artwork will necessarily exhibit all these characteristics at once. His list, he claims, is descriptive, not normative. It only provides "a useful guide" rather than a magic "formula that allows us to crank out an answer to every question about whether something is or is not art." Nonetheless, Dutton insists that if we encountered some mysterious object that, although considered art by some people, does not have any "discernable connection with this established complex of ideas," this can only mean that "the mysterious object is not a new kind of art: it is not art at all."[12]

As noted above, the "universal nature of art" is a key concern for neo-Darwinian scholars of art and culture. They are adamant that art history and aesthetics make sense only if we are able to define art in quantifiable terms—be it along a cluster of eight or twelve or any other number of its allegedly essential characteristics.[13] I have already challenged this logic in my earlier critique of meme theory. Much as culture has no basic units, art has no essential characteristics. This is not to deny that it is possible, and might prove productive in some cases, to *invent* such characteristics for the purpose of statistical research and data collection, as Dutton and others have done. Yet contrary to their claim, this quantitative approach will not uncover the *essence* or *nature* of art—although it will, I fear, serve to cover up the distinctly aesthetic traits and means by which particular artworks disrupt statistical knowledge and resist scientific classification. "The conception that objects have fixed and unalterable values is precisely the prejudice from which art emancipates us," John Dewey put it back in 1931.[14] The task of art, Adorno likewise insisted, is precisely "to make things of which we do not know what they are."[15] Put differently, if indeed we should ever encounter the kind of mysterious object evoked by Dutton above—that is, an object that defies any and all hitherto established characteristics of art—its existence would actually serve to prove the exact opposite of what Dutton thinks it would. Given the paradoxical nature of modern art and the desperate search for the "new" in today's art market, any attempt to prove that this mysterious object is *anything but art* would inevitably *transform this object into art*. This paradox is easy to verify: just name an object you think cannot possibly be marketed as art (e.g., urinals, soup cans, cleaning materials, human feces), and you will find out that it already has. Markets are impulsive, not rational, and the art market in particular reflects the fact that art will always defy the logic that tries to define it.

As these comments indicate, I believe that the relevance of evolutionary aesthetics depends entirely on the perspective and timeframe we chose. Its relevance increases the further we look back in time to the prehistoric cultural roots of our species. There is good reason to support Ellen Dissanayake's anthropological view that the biological core of art is "making things special." Art, she claims, is best understood not as a final product, but as a human behavior rooted in religious practices, rituals, play, and many other forms of ancestral cooperation and social engagements that served to increase the chances of (individual or group) survival during the EEA: "Groups whose individual members had the tendency to make things special would have had more unifying ritual ceremonies, and thus these individuals and groups would have survived better than individuals and groups that did not."[16] Similarly, Kathryn Coe's claim that artistic behavior can be traced back to the mother–child dyad and family behavior strikes me as reasonable and worthwhile.[17]

At the same time, however, the explanatory power of evolutionary aesthetics steadily decreases as we move forward in time toward cultural modernity and present-day art. This becomes evident when anthropologists like Dissanayake and Coe feel compelled to challenge or replace, rather than complement, modern aesthetic theory. Their totalizing view all too quickly succumbs to the empty revolutionary rhetoric of today's biologism. Thus Dissanayake, too, proclaims that to "adopt a human nature (or 'adaptationist') point of view is a revolution in worldview for the humanities that can be likened to replacing a geocentric with a heliocentric perspective in cosmology."[18] Hardly. A central concern of this book is to demonstrate that such patronizing claims are both overblown and counterproductive for a meaningful collaboration between the sciences and the humanities. Dissanayake's analogy is fundamentally skewed, because neither art nor aesthetics are scientific disciplines defined by fundamental laws and first principles. If they were, they would lose their proper purpose and unique perspective for the study of cultural phenomena.

Viewed in their proper context and from the right perspective, however, I see no contradiction between Dissanayake's anthropological understanding of art as a universal expression of human behavior (i.e., art is an adaptation) and Dickie's art-historical view of art as a specific expression of cultural modernity (i.e., art is whatever people say it is). The two propositions are not mutually exclusive, and both are plausible, in my view. The difference between them comes down to the historical perspective and evolutionary time frame used by the respective disciplines. If we extend our window to the later days of the Pleistocene, we can easily agree with Dissanayake that "making special" is, indeed, constitutive of being human—as long as this aesthetic behavior is defined in sufficiently abstract terms to incorporate the infinite ways that humans make things "special." But this insight has little, if any, bearing on the meaning of a particular work of art or the historical significance of modern art movements.

In the following, I shall argue that the major weakness of evolutionary aesthetics, apart from its generic disinterest in close readings, is its steadfast refusal to recognize the historical and conceptual complexity of modern aesthetics. As mentioned throughout this book, it is crucial to distinguish between the philosophy of art and aesthetic theory. Although there is significant overlap between these two fields, they do not necessarily share the same object of study. Unlike the philosophy of art, aesthetic theory not only engages the history of man-made cultural artifacts, but also raises questions about the realm of natural beauty that have little, if anything, to do with the nature of art and individual art objects: "How we respond to an object to which we ascribe aesthetic value may thus be significantly different from the way we attend to art," Peter Gilgen rightly notes.[19]

The failure of evolutionary aesthetics to heed this distinction between art and aesthetics helps explain both its revolutionary rhetoric and its conceptually muddled analyses of human culture and artifacts. Among the few scientists in contemporary evolutionary aesthetics who honor this crucial distinction is Anjan Chatterjee, and it is hardly accidental that his book is one of the best monographs on the subject to date.[20] A practicing neurologist doing scientific research in evolutionary psychology, Chatterjee explicitly rejects Dutton's claim that the human mind features an innate art instinct. Citing relevant neuroscientific studies, Chatterjee emphasizes that "there is no art module in the brain. Rather, our subjective experience when we encounter art is cobbled together from bits and pieces of the brain that are used to do other things."[21] Dutton's theory is far too simplistic, because it is based exclusively on the massive modularity thesis of EP, which fails to account for cultural evolution and the plasticity of the human brain (see chapter 3).

The epistemological consequences of EP's excessive reductionism are evident throughout *The Art Instinct*. Dutton's claim, for example, that "aesthetic theory is merely [the] handmaiden" of the arts is about as meaningful as declaring mathematics the handmaiden of engineering or biology the handmaiden of the pharmaceutical industry.[22] Statements like these betray an alarmingly naïve view of human knowledge—as if it were a monolithic, unidirectional, and hierarchically organized structure that extended steadily from the most foundational and most rigorous disciplines at the bottom to the most speculative and least rigorous disciplines at the top. Academic research, however, does not operate this way. Human knowledge and understanding do not evolve in linear fashion from the natural sciences via the social sciences all the way to the humanities. Although the lower levels of our knowledge system restrict its degrees of freedom at the upper levels, the bottom does not determine what happens at the top, and science can neither determine nor predict the knowledge produced in and for particular local environments as studied by the humanities. Instead, human knowledge depends on and produces a complex array of feedbacks, contradictions, and redundancies both within and between academic disciplines, none of which is subservient to any other. Hence it makes no sense to claim that one discipline is more important or more rigorous or more objective than any other. Absolutist claims of this kind are ruined by the inevitable norm-circularity that sustains academic disciplines and enables the specific knowledge they produce.

For all its flaws, structuralist theory of the 1950s had a far better grasp of academic research and knowledge production than does evolutionary aesthetics today. Take, for example, the work of the great Northrop Frye. Similar to literary Darwinists, Frye criticized the lack of scientific rigor in contemporary literary studies, which was dominated by the New Critics' focus on close textual readings at the time. The New Critics, however, had actually developed their methodology precisely in order to formalize what they considered to be the *science* of literary interpretation. Their overarching goal, too, had been to provide *a scientific method* for the "objective criticism of the work of art."[23] And they were not the only ones trying. Throughout the modern history of literary studies, scholars again and again have sought to strengthen the legitimacy of their discipline by propping up its scientific credentials. The proposed methods and approaches range from Wilhelm Dilthey's philosophical hermeneutics at the end of the nineteenth century all the way to Franco Moretti's digital humanities at the beginning of the twenty-first, with lots of distinct variants in between.[24]

In light of this history, one might say that literary studies suffer less from a *lack* of scientific methods than from an *abundance* of such methods. That, precisely, was Frye's conclusion in his *Anatomy of Criticism* from 1951. As his title suggests, Frye argued that the study of literature was in dire need of *one* central hypothesis to inform and guide the scientific analysis of texts. He singled out evolutionary theory as the ideal paradigm for the kind of scientific principle that literary criticism should seek to emulate: "I suggest that what is at present missing in literary criticism is a co-ordinating principle," Frye wrote, "a central hypothesis which, like the theory of evolution in biology, will see the phenomena it deals with as parts of a whole."[25] After more than half a century, evolutionary aesthetics has finally taken Frye at his word. It redeems his scientific aspiration in the most literal sense possible—namely, by establishing evolutionary theory itself as the central axiom for the study of literature.

Yet, unlike biologism today, Frye also recognized that the study of art and literature requires a broad variety of distinct perspectives necessary to address its wide-ranging spectrum of objects and procedures. Critical readings of literary texts include etymological analyses of single words no less than archetypical descriptions of centuries-old narrative patterns across different genres—and everything in between, as Frye pointed out in the *Anatomy of Criticism*. The nuanced perspectivalism Frye developed throughout his work is as relevant today as it was then—not only for literary studies, but for bioaesthetics overall:

> This inductive movement towards the archetype is a process of backing up, as it were, from structural analysis, as we back up from a painting if we want to see composition instead of brushwork. In the foreground of the grave-digger scene in *Hamlet*, for example, is an intricate verbal texture, ranging from the puns of the first clown to the *danse macabre* of the Yorick soliloquy, which we study in the printed text. One step back, and we are in the Wilson Knight and Spurgeon group of critics, listening to the steady rain of images of corruption and decay. . . . But after all, we say, we are forgetting the genre: Hamlet is a play, and an Elizabethan play. So we take another step back into the Stoll and Shaw group and see the scene conventionally as part of its dramatic context. One more step, and we can begin to glimpse the archetype of the scene.[26]

To be sure, Frye himself, at times, lost sight of this fine-tuned analytical spectrum for literary studies, and his fervent search for Jungian archetypes and ancient myths often led him to ignore the aesthetic complexity of singular texts in favor of their commonly shared topoi and overall structure.[27] Yet Frye never once questioned the legitimacy of diverse approaches both within and between academic fields, and he was a strong advocate for their peaceful coexistence. Biologism, however, is not. In his zeal to establish evolutionary theory as the central hypothesis of literary criticism, Joseph Carroll, the founder of literary Darwinism, claims that he and his colleagues must face a harsh reality: either their opponents in the humanities will "continue to prohibit their students from taking up this line of investigation" or, alternatively, literary Darwinism will finally be able to "absorb and supplant every other form of literary study [and] will assimilate all the existing concepts in literary study."[28] According to this logic, evolutionary aesthetics cannot coexist alongside other methodological approaches or epistemological paradigms. Literary Darwinism either rules or it must die trying, end of story.

Carroll's logic illustrates the fact that evolutionary aesthetics essentially revolves around questions of survival and reproduction, as we shall see later on. Yet to suggest that not only living creatures, but academic paradigms, too, are engaged in an existential struggle of life and death is more than unfortunate. It is deeply troubling, because it betrays both a disturbing insensitivity to the history and rhetoric of social Darwinism as well as a profound misconception of the basic epistemological principles that govern contemporary science. Ideas and disciplines certainly compete for intellectual and institutional resources across the board, but they are not locked into some mortal combat for the survival of the fittest. Carroll's claim displays a lack of familiarity with the methodological pluralism in the humanities that seems less informed by science than by ignorance. Despite its claims to pursue the consilience of knowledge, evolutionary aesthetics, in fact, has been remarkably aloof about the particular kinds of research currently done in the humanities. For the most part, this aloofness stems from biologism's overly polemical reaction against postmodern theory. Brett Cooke, for example, considers deconstruction tantamount to a "mass delusion" that stipulates the total "meaninglessness of art."[29] Absurd statements like this abound in evolutionary aesthetics. Literary Darwinists in particular have failed to provide a differentiated analysis of the various methodologies and theoretical approaches currently employed in literary studies. Instead, they reduce all of them to a single overarching paradigm defined solely by its alleged resistance to evolutionary theory and human nature.

In 2010, for example, Brian Boyd, Joseph Carroll, and Jonathan Gottschall claimed to "provide an overview of the literary theory of the twentieth century," even though they spend less than two pages discussing both the traditional humanist and the post-structuralist strands of twentieth-century criticism. Their list includes "impressionistic and improvisatory commentary," on the humanist side, as well as "Marxist social theory, Freudian psychology, Jungian psychology, phenomenological metaphysics, deconstructive linguistic philosophy, and feminist gender theory," on the post-structuralist side.[30] What supposedly unites these disparate theoretical approaches, according to Boyd, Carroll, and Gottschall, is their conceptual indebtedness to Foucault's philosophy, which has become the "dominant conceptual matrix of literary study" since the 1980s. In another collaborative essay from 2012, the authors go further still, declaring that the approaches listed above can effectively be *reduced* to one version or another of what they call "Foucauldian cultural critique."[31]

I shall not dwell on the absurdity of this claim.[32] My main concern here is another, equally baffling claim that neither Foucault nor any of the theoretical approaches he supposedly inspired "have come to terms with the reality of an evolved and adapted human nature."[33] Foucault, of course, was a student of George Canguilhem, the eminent French historian of science, and intimately familiar with the history of biology and modern evolutionary theory. Indeed, Foucault's dialogue with the life sciences extends throughout his major works starting in the 1960s. What else it at stake in Foucault's discussion of *Madness and Civilization* or the *History of Sexuality* if not "human nature"—even though the term itself is rarely used? Toward the end of his career and to the ire of many of his followers, Foucault even dedicated his final lectures at the Collège de France to "The Hermeneutics of the Self," a topic so deeply invested in the question of human nature and subjectivity that his friend Gilles Deleuze felt compelled to reassure friends and foes alike of Foucault's steadfast commitment to a historical–materialist rather than scientific–idealist understanding of what it means to be human.[34]

In short, the polemical claim that Foucault and other cultural critics today simply ignore or do not "come to terms with human nature" says little about Foucault yet lots about the extremely reductionist view of human nature that informs today's biologism. For the only terms that, in the eyes of literary Darwinists, adequately reflect "the reality of an evolved and adapted human nature" are, of course, the sociobiological ideas developed by neo-Darwinian theory, yet increasingly rejected across the humanities since 1975. And since Foucault did not address human nature *in these terms,* biologism concludes that he—and the humanities at large—must have failed to engage the topic *in any terms* at all. But that is not the case. The problem is that biologism fails to distinguish between "human nature" and "the nature of being human" (see chapter 2). The question of what it means to be human remains a crucial concern for many humanities scholars, in particular those who are interested in the "new materialisms" that influence today's "posthumanities."[35] Yet they shy away from using the term "human nature" because it is too restrictive for the kind of research they do. As Boyd, Carroll, and Gottschall point out, to adopt that term is to adopt the following two premises: first, "that differences across [human] cultures and ideologies are dwarfed by similarities," and second, "that our minds have been shaped to provide us with mostly valid information about our world, rather than to operate by arbitrary convention."[36]

These premises, however, are not particularly helpful in many areas or contexts across the humanities. With regard to the first claim, for example, I wonder by what measure, from what perspective, and to what purpose Carroll adjudicates that the differences among human cultures are less significant than their similarities? Surely, an ethnologist who studies the different dialects and cultural practices of small African tribes living in the same environment would hardly consider the cultural differences she observes "dwarfed by the similarities" between these tribes. Otherwise, her study would not make much sense—and, in all likelihood, would not have received the necessary funding to begin with. Similarly, a literary scholar who examines the distinct use of literary tropes and metaphors in mid-twentieth-century American poetry could hardly care less about the fact that language as such is an adaptation dating back to the EEA, because the historical gap that separates human biological evolution from our mid-twentieth-century cultural practices renders this fact entirely irrelevant for a detailed study of the manifold ways in which poetry probes the limits of representational language.

I hope this clarifies why the humanities' critique of "human nature" is anything but "irrational." Nor does it mean that the humanities are "failing to come to terms" with the nature/nurture problem. Biologism's polemic reaction against the alleged "provincialism" of the "Foucauldian paradigm" has, in effect, blinded neo-Darwinian scholars to the manifold ways in which the humanities have started to engage evolutionary theory precisely in areas where biologism remains most deficient so far—feminist theory, above all, but also in political philosophy and media history, not to mention other arts and genres such as architecture, sculpture, or abstract poetry that, unlike Shakespeare and nineteenth-century realist novels, do not lend themselves quite as easily to sociobiological explications of "human nature." The debilitating weakness of evolutionary aesthetics, in my view, is precisely its premise that *all* aspects and manifestations of human art and culture *must submit* to rigorous scientific analysis, and that literary meaning, too, can and should be explained primarily, or even exclusively, in basic evolutionary terms like adaptation, selection, and reproduction. Even the stories we tell, D. S. Wilson claims, have "genelike properties."[37] It follows that anybody who disavows the supreme explanatory power of evolutionary theory is either a postmodern relativist or a denier of human nature—and likely both: "From the postmodern perspective," Carroll laments, "any appeal to 'human nature' would necessarily appear as a delusory reification of a specific cultural formation."[38]

Yes, it would, and for good reasons, too. Postmodernism serves to remind us that the human world is both *real and constructed, both material and conceptual in nature.* And I should hope that neo-Darwinians, too, will concede the fact that the *concept* of human nature has been abused throughout history by war criminals and other perpetrators in order to justify rape, torture, and murder precisely by evoking the allegedly "less-than-human" nature of their victims—from the European conquistadores' treatment of the indigenous peoples of America as "nonhuman" because they were unfamiliar with Christianity and did not speak Spanish or Portuguese, to the United States Constitution of 1787, which defined a slave as counting exactly as "three-fifths of all other Persons," all the way to the Nazis' embrace of social Darwinism and biopolitics that depicted Jews as rats infecting the healthy organism of the German *Volk*, or the Hutu-led government in Rwanda calling for the extermination of the Tutsi "cockroaches" as the only means to protect their biological and cultural identity as a united people. "It is almost impossible," says Tony Davies, "to think of a crime that has not been committed in the name of humanity."[39]

In light of this history, humanities scholars in general, regardless of their disciplines, tend to be suspicious of any scientific or philosophical argument explicitly based on "human nature," and they refuse to share E. O. Wilson's conviction that it will soon "be possible to define human nature with greater precision."[40] Despite these misgivings, however, I venture to claim that few humanities scholars would deny the fact that human beings share an evolutionary history that determines the biological nature of our species. At stake in bioaesthetics, however, is less the scientific effort to define the nature of being human than the methodological question of how best to apply scientific concepts to our historically inflected understanding of art and human culture.[41] Literary Darwinists turn this logic upside down, because they consult literature solely to confirm scientific theories. In their view, the "scientific approach [to] literary analysis" yields rich statistical data about gender relations and reproductive processes that provide empirical evidence for the universality of human nature across space and time.[42] In this chapter, however, I hope to demonstrate that statistical analyses of this kind and the treatment of fictional characters as if they were real living people whose behavior accords with evolutionary principles of generic human nature is not the most productive, and arguably the least interesting way to approach the study of literature, let alone art and culture in general.

A Survey of the Field

In 2008, the American journal *Style* published a special double issue centered on a target article by Joseph Carroll, the founder of literary Darwinism, titled "An Evolutionary Paradigm for Literary Studies."[43] The book-length collection of essays included brief responses to Carroll's work by more than thirty international scholars from numerous disciplines across the arts and sciences as well as Carroll's detailed rejoinder to his critics. The debate stands out because of its broad, interdisciplinary scope and the fact that, unlike previous anthologies, it not only included proponents of evolutionary aesthetics, but gave voice to its critics as well.[44] Even more significant, in my view, is that Carroll tried to provide the first comprehensive overview of this rapidly changing field of evolutionary aesthetics: "As things currently stand," Carroll stated in his target essay, "the use of cognitive psychology in literary study can be located on a spectrum running from deconstruction at one end to evolutionary psychology on the other."[45] On the deconstructive side, Carroll situates scholars associated with "cognitive literary studies" and "cognitive cultural studies" (e.g., Lisa Zunshine, Ellen Spolsky, Alan Richardson, Patrick Colm Hogan); on the EP side of the spectrum, Carroll lists his fellow literary Darwinists (e.g., Brian Boyd, Jonathan Gottschall, John Johnson, Daniel Kruger) as well as film theorists like David Bordwell and Noel Carroll.

As is frequently the case in evolutionary aesthetics, the *Style* debate, too, largely focused on open research questions in evolutionary psychology and cognitive science that have little, if any, direct relevance for the study of art and aesthetics. To the degree that Carroll does address culture and the arts specifically, he remains committed to raising essential questions only—including, of course, the one about the "nature" of art. Carroll deliberates at length about whether we should define art in terms of human *behavior*, as Ellen Dissanayake suggests, or in terms of art's cognitive *function*, as Brian Boyd argues, or, finally, along a numerical *list* of "signal characteristics" allegedly shared by any and all works of art across time and space, as Dennis Dutton claims. And even though most neo-Darwinians agree that, anthropologically speaking, art emerged from ceremonies and rituals that served to increase the social cohesion within and among hominid groups, thereby increasing these individuals' and their groups' chances for survival, some scholars have relativized the survival value of art in comparison to its role during sexual selection and reproduction. The distinction between natural and sexual selection dates back to Darwin himself, of course, but it has regained prominence more recently through the work of Geoffrey Miller.[46] Miller's favored example of how "art" aids sexual selection is the peacock's tail, whose striking features at first glance appear to be maladaptations, because they strain the organism's limited resources and draw attention from predators. Yet the male's colors also send an honest signal to females that this particular specimen can afford this costly extravagance precisely because it has exemplary genes and would make an excellent mating partner. The "artistic" quality of the peacock's tale thus increases the male's chances for genetic reproduction—or so the story goes.[47]

A related question one encounters again and again not only in the *Style* debate but throughout the field of evolutionary aesthetics regards the adaptive or maladaptive *function* of art. Is art, indeed, an adaptation in the strong Darwinian sense, or are art and culture merely nonadaptive by-products of our brain? These are the kind of questions neo-Darwinians love to ponder. The majority of critics (e.g. Carroll, Dissanayake, Dutton, Boyd) list numerous reasons why and how art helped increase our chances for reproduction and survival, yet they readily admit that some of these previously useful adaptations may no longer be useful today.[48] Steven Pinker, on the other hand, famously referred to our enjoyment of music as some "auditory cheesecake" that tickles the senses and enriches our lives but "confers no survival advantage" as such. Although Pinker admits that "some aspects of the arts" may have had an adaptive function back in the EEA, he insists that "most" did not.[49] One should also note that Pinker accused others, and Carroll in particular, of harboring the same flaws and misconceptions about Darwinian theory that, according to Carroll, are prevalent across the humanities today.[50] The fairly sharp exchange between Carroll and Pinker was fueled by their opposing views of orthodox EP and its massive modularity theory, which Pinker still supports yet Carroll has come to consider inadequate in recent years.[51] But the fact that both sides accused the other of a limited understanding of *real* Darwinian theory also testifies to the considerable ideological tension and political struggles *within* biologism, depending on whether scholars are more committed to sociobiology or EP. It reminds us that neo-Darwinian theory, though hegemonic across the two cultures, is not monolithic, but fractured, and that scientific debates often serve a polemical rather than strictly scientific purpose, as I argued at length in my second chapter.

It is not necessary—nor would it be possible—to repeat all the particulars of this first debate about literary Darwinism or the one that followed just a few years later in the journal *Critical Inquiry* between 2010 and 2012. Both debates were productive and are worth reading in their entirety. Carroll's target essay from 2008, however, remains particularly instructive not only with regard to the topics he chose to address, but particularly with regard to those he failed to mention. Thus Carroll does not refer at all to the entire field of neuroaesthetics, nor does he distinguish between the diverse media, genres, and objects (e.g., brains and computers, dance, paintings, music, novels, film—architecture and sculpture, on the other hand, are rarely mentioned at all) collectively studied by neo-Darwinian scholars of human culture. As mentioned above, Carroll is less interested in *differentiating* the various lines of research in evolutionary aesthetics than in *unifying* them under his own banner of "literary Darwinism." This disciplinary rigor, too, is typical for the field at large. From its very beginning dating back to E. O. Wilson's *Sociobiology* in 1975, evolutionary aesthetics has been extremely disciplined, both in the sense of trying to build a strong network of scholarly research (including frequent publications, conferences, and concern for public funding and popular appeal) and in the sense of holding fast to the neo-Darwinian paradigm of scientific reductionism and its rigid notion of human nature. Remember that Dennett likens evolutionary theory to some "universal acid" able to reshape the entire field of human knowledge.[52] E. O. Wilson put it this way: "The central idea of the consilience world view is that all tangible phenomena, from the birth of the stars to the working of social institutions, are based on material processes that are ultimately reducible, however long and tortuous the sequences, to the laws of physics."[53]

I have already discussed the numerous methodological and conceptual problems of this totalizing perspective in previous chapters. Carroll hardly mentions any of them at all. Instead, he explicitly praises Wilson's "vision of a unified total body of knowledge" and reiterates his claim "that science can obtain reliable knowledge about the structure of nature, and that human art, like all other objects in nature, can be reduced to underlying regularities and submitted to scientific understanding." More specifically, literary Darwinists "contend that human nature is both the source and subject of literature," and they believe that evolutionary theory can help us "understand what literature is, what its functions are, and how it works—what it represents, what causes people to produce it and consume it, and why it takes the form it does."[54] This is a tall order. Carroll, in fact, claims that literary Darwinism can or will soon be able to explain pretty much all there is to know about literature. Carroll reiterated this scientist premise in his 2008 essay:

> Like Wilson, they [literary Darwinists] envision an integrated body of knowledge extending in an unbroken chain of material causation from the lowest level of subatomic particles to the highest levels of cultural imagination. And like Wilson, they regard evolutionary biology as the pivotal discipline uniting the hard sciences with the social sciences and the humanities.[55]

Carroll's biologistic version of bioaesthetics is profoundly misleading. First of all, he never seriously engages the ongoing philosophical debate about the irreducibility of biological principles to basic physical laws. If consilience were possible, biology would have never emerged as a scientific discipline in the first place, or it would have been absorbed into physics long ago. Second, Carroll never mentions the wide spectrum of different and often contradictory interpretations of evolutionary theory, among which neo-Darwinism provides only one, albeit influential, version, as we saw in previous chapters. Finally, he simply ignores the fact that human cultural evolution cannot adequately be conceptualized in neo-Darwinian terms, but requires a far more complex methodology and more refined concepts (e.g., social learning, sociogenesis, technogenesis) than either sociobiology or EP was able to provide.

Refusing to consider any of this evidence, Carroll extols Wilson's dictum—genes lead, culture follows—as if it were holy writ. But it is just a working hypothesis, not a scientific fact, because there is no way to verify this assumption empirically. The study of human populations in systems biology leads to the emergence of statistical trends, not scientific laws.[56] System biologists do not observe living people(s); they study mathematical algorithms that simulate population growth in artificial, computer-generated environments. These computer models severely restrict the degrees of freedom of the modeled system—that is, they mathematically reduce the number of relevant parameters so as to reflect some values and not others. Even the processing capacities of today's most advanced mega computers are insufficient to map a complex topological space defined by hundreds or thousands of different parameters at once—the kind of infinitely complex environments that constitute human cultures.

Carroll, however, insists that there "are no real ontological or epistemological barriers separating the humanities and the evolutionary human sciences."[57] He ignores the fact that neo-Darwinian studies of cultural evolution are steeped in statistical analyses of empirical data assembled for particular purposes within specific contexts and limited parameters. This method obviously differs from the humanities' traditional focus on close reading and historical analysis of specific artifacts. A related problem is Carroll's refusal to take seriously the instability and aleatoric behavior of complex systems, where causality does not operate unidirectionally from the bottom up, but also extends from the top down. As argued above, it makes no sense to conceptualize the dynamic relation between natural and cultural evolution in quantitative terms that implicitly or explicitly grant priority to either of the two processes as if they could be studied apart from each other. Hierarchical models of this kind are flawed, because they fail to account for the constitutive epistemological constraints of modern science.[58] Carroll's simplified view of material causality as an "*unbroken* chain" of causation is more than just misleading. It is plain wrong. Emergence theory clearly demonstrates that the behavior of nonlinear complex systems defies the logic of mechanical causation.

Although Carroll has been criticized repeatedly for his Newtonian understanding of post-Newtonian science, he (along with Steven Pinker and other neo-Darwinians) remains defiant: "The argument on 'complexity,'" he rejoins, "is just sophistical drapery arranged to place a decent covering over [critics'] belief in the autonomy of culture."[59] Like Wilson before him, Carroll insists that whoever questions the objective validity of consilience and neo-Darwinian theory is either a cultural constructivist, a postmodern irrationalist, or some other kind of ideologue who subscribes to the delusional belief that "culture is a law unto itself, a transcendent category causally disconnected from any biological antecedent."[60] This undifferentiated, binary juxtaposition of evolutionary versus postmodern aesthetics disregards the fact that some versions of social constructivism are easily compatible with the Darwinian approach, as David Sloan Wilson and Ellen Spolsky have demonstrated.[61] Neuroaesthetics is another case in point. Although based on molecular science, neuroaesthetics, too, emphasizes the "constructive" side of human perception and cognition—the fact that our brains actively construct rather than passively reflect the human environment.

A related problem is that Carroll fails to clarify the specific meaning of the "cultural autonomy" he rejects. Although he acknowledges the existence of feedback loops between biological and cultural evolution, he denies that this structural recursivity has any effect on "the unbroken chain of material events" in science. To defend this claim, Carroll occasionally juxtaposes the consilient worldview of "unified causal explanation" to another, more simplistic model he refers to as "simply unilinear causation—no feedback loops."[62] According to Carroll, the confusion between these two models informs the widespread critique of neo-Darwinian theory as too reductionist and incompatible with complexity theory. But Carroll himself never elaborates or provides specific details regarding this alleged difference between "unified causal explanation" and "simply unilinear causation." Nor is this distinction ever mentioned in any of the scientific literature I read in fields like evolutionary epistemology, chaos and complexity theory, or analytical philosophy. Instead of charging his critics with failing to observe an epistemological distinction he alone claims to be relevant, it might be more helpful if Carroll himself provided a detailed explanation as to why the overall instability and myriad structural effects of feedback loops in complex systems, in his view, support rather than undermine the starkly reductionist model he so vigorously defends.

A similar explanatory gap undermines another conceptual distinction Carroll introduced recently—namely, that between "gene–culture coevolution," on the one hand, and straightforward "cultural evolution," on the other: "The argument for 'cultural evolution,' as distinct from gene–culture coevolution, is that certain cultural figurations ('memes') can work against reproductive success and nonetheless persist within limited parts of a population for a few generations."[63] This statement is misleading because it falsely identifies cultural evolution tout court with Dawkins's problematic meme theory. We have already seen that Boyd and Richerson's model, for example, makes little use of meme theory other than pointing out its limitations. Since Carroll misconstrues cultural evolution as meme theory, he falsely assumes that "cultural evolution in this sense did not drive the evolutionary process that produced key features of an evolved human nature—big brains, linguistic skills, and ultra-sociality," and that, therefore, "we need make no appeal to 'cultural evolution' as a causal force that has 'autonomous' power to direct the course of evolution."[64] That, too, is wrong. Even if we disregarded the question of whether cultural evolution "has 'autonomous' power"—a question that scholars like Monod, Changeux, Tomasello, Boyd and Richerson, and many others have already answered affirmatively—few social scientists today would want to deny the significantly weaker claim that cultural evolution constitutes "*a* causal force . . . to direct the course of evolution."

Take Friedrich Hayek, for example, arguably one of the most influential thinkers in twentieth-century economics: "Once an innate capacity for learning by imitation is acquired," Hayek stated some forty years ago, "the transmission of abilities takes a new form—vastly superior to genetic transmission precisely because it includes transmission of acquired characteristics which genetic transmission does not."[65] Hayek was a founding member of evolutionary economics, a field that tries to reconcile the Lamarckian theory of inherited characteristics with the Darwinian theory of natural selection. The premise is that natural selection does not select for a specific trait acquired during an individual's lifetime, as Lamarck falsely assumed, but instead selects for the organism's innate ability for learning by imitation. As Geoffrey M. Hodgson puts it, natural selection favors traits that enhance the possibility of further accelerated evolution: "Evolvability," Hodgson declares, "is the greatest adaptation of all."[66] Obviously, Hodgson does not deny that biological evolution remains fundamentally important. If, for whatever reason, the human species were no longer able to reproduce and ceased to exist, then human culture and economics would inevitably follow suit.[67] This, however, does not mean that natural selection suffices to explain cultural evolution, much like Newtonian physics is insufficient to explain biological evolution: "The notion of Universal Darwinism itself," Hodgson clarifies, "provides no alternative to a detailed explanation of the particular emergent properties and processes at the social level."[68] Hodgson's repeated reference to emergent properties in complex systems distinguishes his understanding of evolutionary theory from that of literary Darwinists who reject all talk of complexity as empty rhetoric. Unlike Carroll's, Hodgson's model leaves ample space for "the possibility of different kinds of evolutionary processes at different ontological levels."[69]

The numerous epistemological flaws of literary Darwinism culminate in one crucial point—namely, that contrary to Carroll's assumption, scientific progress and the successive accumulation of human knowledge cannot, and must not, be conceptualized in Darwinian terms of natural selection. Although concepts might seem to behave "like" organisms in some respects, both meme theory and evolutionary aesthetics have pursued this misleading analogy to the point of absurdity. As Michael Ruse explains in his detailed discussion of neo-Darwinian theory, that model is based on the false "analogy between the development or evolution of organisms and the supposed development or evolution of (scientific) knowledge." Ruse strongly rejects this analogy as little more "than an insightful metaphor."[70] Gerhard Vollmer puts it succinctly: "Despite all similarities, analogies, and parallels, the evolution of science is not Darwinian."[71] Yet in the eyes of literary Darwinists, evolutionary theory will not only revolutionize the humanities, but also, and more importantly, presents the last hope for "studies in literature and film [to] ensure their survival."[72]

To be clear, I fully agree with Carroll that the humanities should widen their analytical scope and embrace evolutionary and cognitive theory as part of their research program. And for all it its limitations, literary Darwinism has been a trailblazer in demanding that literature departments start taking Darwin seriously, which they should. Yet it is irresponsible, in my view, to suggest that unless literary Darwinism (or evolutionary aesthetics in general) achieves total dominance in literary studies, it will either go extinct or survive only in "its vulgar form." Much as bioaesthetics strives to welcome Darwinian theory into the disciplinary fold of the humanities, it cannot but refuse this totalizing logic of biologism. There is simply too much competition and too little cooperation in Carroll's view of interdisciplinarity. If the humanities were to subscribe to this logic, they would begin to lose their senses, literally. They would no longer strive to be affected by the nonconceptual knowledge and the mind-boggling diversity of human artifacts and cultures.

Literary Darwinism Revisited

Literary Darwinism is currently a field in decline. Its reputation certainly suffered from Jonathan Kramnick's devastating critique "Against Literary Darwinism" in *Critical Inquiry* in 2010 (and in 2012, in Kramnick's response to his critics).[73] Kramnick's title, of course, made implicit reference to the famous article "Against Sociobiology" that had sparked the public debate about E. O. Wilson's *Sociobiology* in 1975. In choosing this title, Kramnick managed to identify sociobiology as the intellectual foundation of literary Darwinism—a lineage Carroll himself is all too happy to embrace—while, at the same time, he aligned himself with Lewontin and Gould as an outspoken critic of neo-Darwinian theory. And just like Lewontin had claimed that there was absolutely no scientific basis for Wilson's theory, Kramnick's main point was that literary Darwinism had failed to provide sufficient evidence for *its* central thesis, according to which literature and storytelling are evolutionary adaptations constitutive of human nature:

> Literary Darwinism has a difficult time finding a place for literary forms in the story of adaptation under selection pressure. At the same time, it is committed to the proposition that literature must have helped us to become the species we are. The result of this curious imbalance is that literature simply is about who we are in a relatively straightforward and uplifting way.[74]

I think Kramnick's critique is spot-on. Unable to substantiate their claim in scientific terms, literary Darwinists typically resort to a purely thematic treatment of the subject—as if their thematic exposition of traditional gender roles and sexual relations depicted in fictional stories were able to prove the scientific claim that such relations exist due to universal predispositions we inherited from the Pleistocene, or the claim that the *telling of such stories* is itself an adaptation (or must have been adaptive during the EEA). This analytical confusion between scientific studies and fictional stories amounts to a categorical error about the modality of storytelling. Unlike real people, fictional characters neither evolve nor adapt; they are the creation of their authors. Literary Darwinism is based on surreptitious concepts *(erschlichene Begriffe),* as Kant would say: it improperly transfers scientific concepts from one discipline (biology) to another (literary studies), and thus from the real material world into the world of fiction where they no longer make any sense.

As mentioned above, I am agnostic about whether the art of storytelling is an adaptation, because I do not think literary studies hinges on answering the question either way. Obviously, the emergence of language itself was an adaptation, and it strikes me that artistic behavior in the broad anthropological sense of "making special" was an adaptation, too, though it probably is no longer adaptive today. Beyond that, the question about storytelling as adaptation becomes both increasingly difficult to answer and increasingly irrelevant to bioaesthetics. Instead of reexamining that question, I want to focus on the kind of readings literary Darwinism has been able to produce. I suspect that the ongoing dissolution of the movement—in addition to evolutionary scholars across the disciplines, even former key members like Brian Boyd and Jonathan Gottschall have distanced themselves from Carroll's paradigm—is due less to its inability to prove that storytelling is an adaptation than to its predictable and trivial interpretations of literary texts.[75] In order to distinguish biologism from bioaesthetics, we need to better understand how and why literary Darwinism failed not only in scientific terms, but in literary terms as well.

Carroll has repeatedly argued that the "aesthetic value" of literary texts can be evaluated or "grouped into three main categories: the significance of the represented action, the mind and temper of the artist, and formal aesthetic organization."[76] Most literary scholars will surely recognize these three categories as variants of our common analytical distinction between the reader, the author, and the text. So, how does literary Darwinism think through and engage these three categories in their interpretation of literary texts? Let me start with the writer—"the mind and temper of the artist." The author is by far the most important analytical category for literary Darwinism and explicitly acknowledged as such by Boyd and Carroll. Indeed, if the linguistic turn in the 1960s was characterized by "the death of the author," then literary Darwinism is best characterized as the effort to bring him back to life. "The resurrection of the author," I submit, is the central feature and main objective of this movement.

Bioaesthetics, in my view, should welcome a return to authorship. It is a necessary and long-overdue corrective to the fetishization of language in (post-)structuralist theory, the upshot of which had been to minimize the relevance of authors and readers alike, as I will argue below.[77] In literary studies today, authors matter, no less than texts or readers do. Hence I much appreciate Boyd's work on Theodor Seuss Geisel (aka "Dr. Seuss"), not least because it challenges the blatant disregard for nontraditional modes of storytelling (e.g. cartoons, graphic novels, digital animation, creative nonfiction) that continues to be the norm in most literature departments across the country and abroad. To the degree that author-based studies can help expand the academy's narrow-mindedness about the kind of "literature" worth its while, I am all for it.[78] Unfortunately, however, that is not the norm. In their attempt to resurrect the author, literary Darwinists all too often end up being conservative rather than revolutionary. They overwhelmingly turn to canonical works by Shakespeare and nineteenth-century realism as opposed to less traditional and more marginalized periods or genres. And given their overriding concern with the nature of art and its adaptive qualities, literary Darwinists also tend to ignore the aesthetic specificity of individual artworks in favor of common features and universal patterns across the board.

A good example is Carroll's analysis of five nineteenth-century British novels in an essay from 1994. Although Carroll explicitly mentions the three literary categories outlined above, his main objective is to let these novels “serve as a central point of reference for our understanding of human nature.” His two-page comment on Charlotte Brontë’s *Villette* concludes as follows: “The narrator’s relation to her subject in *Villette* is fairly complex, but there is little uncertainty about the total structure of meaning.” The female protagonist, Carroll goes on, “yearns for intimacy, seeks to sustain her dignity, had to come to terms with diminished expectations, and then ultimately must renounce even those expectations. She defines her own position in relation to a normative model of romance, but she also provides in herself a model for the affirmation of personal integrity in the face of insuperable obstacles to happiness.” I think it is fair to say that Carroll is not particularly interested in a close reading or structural analysis of Brontë’s text. His main concern is to emphasize the cognitive intelligibility of the fictional world presented in the novel, an intelligibility that applies to all people, real or imagined, as members of the human species: “The emotional development in the narrator’s point of view follows a clear causal trajectory, and the meaning of that trajectory is available to both the narrator and the reader.”[79]

In the eyes of literary Darwinists, the meaning of the text is plain to see and easy to access. It is a surface phenomenon that does not require much hermeneutic skill or deep structural analysis, because the lines of communication between literary authors, fictional characters, and their readers across time and space are always open, with only few glitches along the way, most of which occur at the molecular level when human nature is distorted in one way or another, as we shall see below. Otherwise, the essential meaning of texts will inevitably emerge at the end of the story, because all three (authors, readers, and characters) share the same mental architecture inherited from the Pleistocene era, and all are rewarded with increased fitness for their continued engagement with literary texts, as Carroll points out in his reading of Dickens’s *David Copperfield*: “By nurturing and cultivating his own individual identity through his literary imagination, he enables himself to adapt successfully to this world. He directly enhances his own fitness as a human being, and in doing so he demonstrates the kind of adaptive advantage that can be conferred by literature.”[80]

The main objective of Carroll’s literary paradigm is to confirm that literature is adaptive and enhances our overall fitness as human beings. Authors invent fictional characters to reflect on their (own) human nature—“The central artistic purpose in *Dorian Gray* is to articulate the anguish in the depth of Wilde’s own identity”[81]—and this authorial message, it turns out, is precisely what readers want to hear: "Writers rule, but only because they provide their subjects with what the subjects want."[82] Yet this logic *undermines* the crucial distinction between the author or writer *of a text,* the fictional characters, narrators, and other imagined beings *within a text,* and the multitude of historical readers *outside the text.* These distinctions make good sense, because writers and readers are real living people, whereas characters and narrators are neither living nor people at all. Literary characters have no bodies, no brains, no genes, and nothing to their name but the series of letters that, literally, spell them out. J-a-n-e E-y-r-e: here she is, Brontë's heroine, in black and white—or in flesh and blood, if you choose to look at her through the eyes of literary Darwinists.

Fictional characters, indeed, readily spring to life in evolutionary interpretations of literary texts. In the following passage, Jonathan Gottschall describes his personal experience of reading a passage in Nathaniel Philbrick's *In the Heart of the Sea,* an experience supposedly shared by all readers anywhere, at any time: "When you read the Philbrick scene, it came alive in your mind. Let me ask you, what did Captain Coffin look like? Was he young or old? Did he wear a tricorn hat or a floppy-brimmed deal? What color was his coat? What color was his beard?" In raising these questions, Gottschall hopes to support his key thesis that "human minds yield helplessly to the suction of story" because "we just can't resist the gravity of alternate worlds."[83] I must confess that while reading fiction, I rarely have any of the visions Gottschall claims I do or ought to have. This might be in part because I am not a native English speaker and often find myself at odds with the English language. But I also find myself wondering about how people learn to read in the first place. Rather than visualizing the flesh and blood of Captain Coffin in her mind's eye, as Gottschall has it, the apprentice reader must first and foremost learn to decipher the proper shape and meaning of those black marks on white paper we call words. The German media theorist Friedrich Kittler has shown that the visualization of literary texts extoled by Gottschall and literary Darwinism as innate to human nature emerged instead from specific reading and writing practices developed and institutionalized across western Europe around 1800.[84] According to Kittler, the kind of mental images we have come to associate with the process of reading are neither natural nor prehistoric. They are a product of modern cultural evolution and due to particular learning techniques and media–technological inventions over the last two hundred years.

I do not mean to belittle the supreme fictional quality of the stories Gottschall tells, but the point must be made: his theory of what supposedly goes on in the mind of the reader is not *de*scriptive, but *pre*scriptive. It is a normative construct informed by sociobiology and orthodox EP, according to which "the reader's" mind is controlled by the mental architecture inherited from the Pleistocene era some two hundred thousand years ago. Yet in making this argument, Gottschall ignores the far more relevant institutionalization of reading and writing techniques during the Enlightenment that continue to shape readers' visualization of narrative texts. And a key characteristic of literary modernism around 1900 was precisely to reject some of these techniques and to emphasize the materiality of language instead. Let me introduce you to Anna Blume, a lovely lady portrayed by the Dadaist writer Kurt Schwitters in 1919:

To Anna Blume

Oh You, beloved of my 27 senses, I love Ya!
You, Yours, You to you, I to You, You to me,——we?
This, by the way, does not belong here!

Who are You, uncounted womanhood, You are, are You?
People say that You were.
Let them talk, they do not know where the church bell rings.
[. . .]

Winning question:
(1) Anna Blume has a bird
(2) Anna Blume is red
(3) Which color is the bird?
[. . .]

Anna Blume, Anna, A———N———N———A!
I drizzle Your name.
Your name drips like soft cow's fat.
Do You know, Anna, do You know it already,
One can read You backwards as well.
And You, You most glorious of all,
You are the same from back or front:
A———N———N———A.
[. . .][85]

I hope Gottschall would agree that it is rather difficult for any reader to picture Anna Blume in her mind. The meaning of this poem remains unaffected by mental imagery of any kind, and it is entirely irrelevant whether readers believe that Anna Blume does, indeed, have a bird.[86] What matters instead is that Anna spells the same forwards as backwards, and that the essence of her being consists of only two letters that can be drizzled on the page in much the same way that Jackson Pollack, decades later, will pour color on canvas.

We have already arrived at the third and final analytical category I want to examine in the context of literary Darwinism—namely, the reader. For much of twentieth-century aesthetics, I would argue, the reader has remained a stepchild of literary criticism and a vastly undertheorized category in the art of interpretation. Most scholars have shown little or no interest in actual reading processes and physically embodied readers as opposed to the "implied reader" supposedly constituted by the text. Take, for example, Roland Barthes's famous essay "The Death of the Author," written almost fifty years ago in recognition of the fact that the linguistic turn had come to dominate—or had "revolutionized"—literary theory at the time. The liquidation of the author was all the rage in revolutionary Paris in 1968. It is high time, Barthes wrote, "to substitute language itself for the person who until then had been supposed to be its owner," because "it is language which speaks, not the author." Only by ridding itself of the authoritarian notion of the "author," Barthes claimed, would literary theory finally be able to emancipate the reader from the shackles of literary conventions—for "a text's unity lies not in its origin but in its destination." Hence the famous conclusion of Barthes's essay: "The birth of the reader must be at the cost of the death of the Author."[87]

His vitalist rhetoric notwithstanding, the kind of reader Barthes had in mind, surprisingly, was neither born nor alive at all. The reader, Barthes insists, "cannot any longer be personal: the reader is without history, biography, psychology; he is simply that *someone* who holds together in a single field all the traces by which the written text is constituted."[88] This reader, "that *someone*" Barthes conjures at the end of his essay—who is that supposed to be? Who in their right mind could possibly hold together "all the traces" that constitute a written text? And if such a reader were to exist, would her control over the text not be ten times more totalizing than that of any author in history? Barthes's reader, I submit, is not *someone*, but *no one*. It is a mere ghost in the machination of linguistic patterns, a structural effect meant to guarantee the continued circulation of language and the production of ever more *discourse*. Bioaesthetics, in my view, can easily do without this structuralist fantasy of mindless and disembodied readers.

Barthes's disinterest in actual readers is typical of much of twentieth-century literary theory, including structuralism, post-structuralism, Marxism, and psychoanalysis. A few exceptions notwithstanding—in fields like phenomenology, feminism, and post-colonialism—literary scholars have shown little or no interest in examining the role of living and breathing readers, and they did so for the very same reason that Barthes did, too: because in their mind, it makes little difference who the reader is or what she does. The reader has nothing to say on the matter of meaning, literally, because the "implied" reader is expected to submit to the scholar's method of interpretation rather than rise to the challenge or emerge as its equal. Thus, according to (post-)structuralism and deconstruction, "the reader" in constituted by textual structures and the being of language; according to Marxism and psychoanalysis, "the reader" remains subject to bourgeois ideology and the working of the unconscious; and according to evolutionary aesthetics, the response of "the reader" is determined by the architecture of her mind inherited from the Pleistocene. In all of these models, the reader is nothing more than a puppet on a string. She exists only as a necessary but hapless analytical category called on to confirm the kind of literary analysis typified in the model. The sole function of this disembodied reader is to ventriloquize the speech of language or ideology, the unconscious or the genetic code. Other than that, the reader has no voice of her own.

At first sight, literary Darwinism appears to make an important contribution in that regard. Like many of his colleagues, Carroll often insists that the "reader is an integral part of the larger communication event, and thus part of the total meaning situation" of literary works.[89] Indeed, the most distinctive trait of literary Darwinists is their empiricism: they collect statistical data from actual readers about what and how they read. That is a good idea, and one likely to become increasingly important as the digital humanities continue to expand. Like most scholars interested in bioaesthetics, I firmly believe that empirical data has an important role to play in the study of literature and culture—as long as the authors of such studies recognize, first, that the resulting data still require skillful interpretative analysis to become meaningful and, second, that the significance of this interpretation depends largely on the specific questions and issues that orient the overall study. The last point in particular addresses a core problem for literary Darwinists. Their sense of art and literary meaning is hamstrung by their central premise that "the ultimate purpose and origin of art" amounts to the "manipulation of the behavior of others in order to gain reproductive advantage."[90] Their main goal, both in their quantitative and qualitative studies of literature, is usually *not* to probe the imaginary power and interpretative skills of the readers they survey, nor do they seek to demonstrate the aesthetic quality or unique characteristics of single works. Instead, their method is designed to demonstrate the ultimate explanatory power of neo-Darwinian theory and to promote its unfettered reign across all fields in the humanities.[91] This explains why literary Darwinists have so little to say about the formal–stylistic aspects of individual texts: their method of interpretation aims at standardization, not differentiation. Unification, not pluralism, is their methodological goal for the future of the humanities. Hence any text they read sooner or later ends up testifying to inborn gender differences and the importance of human survival and reproduction. There are numerous examples to illustrate this paradigm, and they are all equally breathtaking.[92] One of my personal favorites is David and Nanelle Barash's interpretation of Gustave Flaubert's *Madame Bovary*:

> Biologists understand that a major reason why Emma wanted sex with Rodolphe, Léon, and the marquis (the last unconsummated) was because deep inside (in the DNA of her brain) she heard a subliminal Darwinian whisper that tickled her ovaries, even though she may not have acknowledged it and would likely have even acted consciously against such an outcome. Smart women sometimes really do make foolish choices, and a whiff of Darwin enables us to glimpse some of the reasons why.[93]

The reason why this is an exceedingly bad reading of Flaubert is that Emma does not have ovaries. Madame Bovary is not a real person. She is a fictional character in a nineteenth-century French realist novel. The reason we think of her as being real (and having ovaries) is that when we listen to stories or read a book, we tacitly accept the literary conventions that govern the diegetic world of the narrative, which in this case happens to be realism, a genre that assumes a mimetic relationship between the real world outside and the fictional world within the text. But mimesis is neither a necessary nor a coherent element of literary realism, let alone of literature in general, and many narrative forms deliberately challenge or undermine that convention. Kafka's realism, for one, is such that people can suddenly find themselves transformed into a gigantic bug, and the reality described by Edgar Allan Poe is but a dream within a dream . . .

My obvious point is that fictional worlds are not subject to natural laws or biological evolution. Hence ravens can speak and pigs can fly, if authors make it so—and they do: in gothic novels or sci-fi stories, in tales of horror and other visions of the fantastic, the uncanny and the surreal evoked in literary texts over the last few hundred years alone. Once we move beyond nineteenth-century realism and apply literary Darwinism to other periods and genres, its limitations become blatantly obvious. We met Anna Blume above—should we assume that she, too, has ovaries? "Rose is a rose is a rose is a rose," Gertrude Stein insists in one of her best-known poems—does Rose have ovaries?[94] And does Frankenstein's monster have DNA that whispers to his brain? Is that why he chose to take revenge on Victor by killing Elizabeth on their wedding night? I do not think so. More important, I do not think these are good questions to ask or think too much about. Life is short, and literature has far more interesting things to tell us about who we are.

Not so for literary Darwinists. In their mind, the history of literature ultimately comes down to ever-new stories about the importance of human survival and sexual reproduction.[95]

Let me support this claim further with reference to an online questionnaire that Carroll and his colleagues sent to professional readers of literature a few years ago. Their goal was to collect data about these readers' emotional responses to two thousand fictional characters that appear in two hundred canonical British novels of the nineteenth and early twentieth centuries. The researchers drew two overall conclusions from the resulting data, both of which, predictably, reflect the authors' overriding concern with sex and gender differences. Their first conclusion regards the frequent representation of personal strife and conflict in these novels. Although the authors concede that these conflicts "could be conceived as competing value structures" unrelated to biological sex, their final interpretation of the data suggests instead that the persistence of these conflicts must "ultimately be attributed to competing reproductive interests" between women and men.[96] This is unsurprising. We have already seen that literary Darwinists fail to distinguish between real people and fictional characters. Given their premise that gender relations in the real world are dominated by the competing reproductive interests between men and women—a central tenet of neo-Darwinian theory and explicitly endorsed by the authors of this study—it follows that innate gender difference will appear decisive to them in the fictional world of British realist novels as well. This conclusion merely restates the premise of their study. It confirms the authors' belief that sociobiology rules in all possible worlds, be they real or imagined.

The authors' second conclusion seems more interesting. It takes into account the fact that most queried readers, male or female, reacted less to the novels' depiction of gender conflicts than to the depiction of likeable characters (protagonists) and unlikable characters (antagonists), irrespective of their sex. The survey, in other words, confirmed that both male and female readers despise Heathcliff and love Cathy. This finding seems to pose a problem for literary Darwinism, the authors submit, because it appears to contradict (or at least fails to confirm) the Darwinian claim that men and women have inherently different natural dispositions regarding sexual relations and reproduction. For their study shows that regardless of these innate differences, both male and female readers are united in scorning the villains and embracing the heroes in British novels. It follows that men and women must share one overriding goal that dwarfs the inherent differences between them: "The subordination of sex to agonistic role assignments . . . suggests that in these novels conflict between the sexes is subordinated to their shared and complimentary interests."[97]

What are these shared interests of men and women? They are to maintain a stable social order conducive to the safe rearing of their offspring. Members of both sexes are equally interested in the preservation of a healthy community able to protect and facilitate the reproductive process of its members. "In Heathcliff," Carroll states in his interpretation of Emily Brontë's *Wuthering Heights,* "human nature has been stunted and deformed."[98] Hence nobody, neither males nor females inside or outside the text, likes him. His pathological nature makes Heathcliff a threat to any society, fictional or historical, because his presence destabilizes the natural process of biological reproduction that unites human beings in both narrative texts and the real world alike. The main purpose behind the fictional depiction of interpersonal conflicts and agonistic social structures prevalent in nineteenth-century British novels, Carroll and colleagues argue, is to expose those who, like Heathcliff, pose a threat to the community due to their deformed human nature; and this serves to remind both male and female readers that regardless of their inherently different natural disposition vis-à-vis the process of biological reproduction, they still share the far more important goal to identify and isolate these villains in order to prevent their deformed human nature from being reproduced and jeopardizing the well-being of everybody else: "Agonistic structure thus functions as a cultural technology that extends an adaptive social process across social groups larger than the band or tribe. It is a medium both for gene–culture co-evolution and for natural selection at the level of social groups. It is, in other words, an adaptively functional feature of human nature."[99]

I am not quite sure what an "adaptively functional feature" is exactly, but I interpret the last sentence to mean that the agonistic structure that characterizes nineteenth-century British novels is an adaptation. Its evolutionary function is to increase the social cohesion and sexual reproduction among healthy—that is, neither stunted nor deformed—members of the human community that exist both inside and outside the text. What is so remarkable about this conclusion is that it moves beyond the more general claim that the art of storytelling is an adaptation and still remains adaptive today. This study actually claims that one *characteristic trait of one particular literary genre* (i.e., the frequent depiction of personal and social strife in British realist novels) *is itself the result of an adaptation*—a totally specious claim that lacks any scientific foundation whatsoever. A singular and extremely limited survey of contemporary readers' responses to British novels falls way short of the methodological framework and massive empirical data necessary to support Carroll and colleagues' theory. In order to make any sense of it at all, the minimum requirement would be to conceptualize a specific environment either within or outside literary texts able to select for this particular trait over a sustained period of time. Put differently, Carroll and colleagues would have to determine first that the popularity of British realist novels in the nineteenth century was due primarily or exclusively to its feature of antagonistic structure (as opposed to, say, its beautiful descriptions of natural landscapes or bourgeois interiors), and second that the main or sole reason readers enjoyed that particular feature was the adaptive advantage it provided for readers and their communities (as opposed to, say, mere *Schadenfreude* or readers' vicarious interest in feuds of any kind regardless of their adaptive function). But we already know from the previous chapter that it is simply impossible to trace the evolutionary history of any particular cultural unit empirically, and the same is true of the "antagonistic structure" of nineteenth-century British novels. Much as the evolutionary principle of natural selection does not regulate today's art market, the process of adaptation cannot possibly explain the relative frequency of particular topoi or formal structures encountered for limited periods of time in particular literary genres. Any claim to the contrary is pure speculation.

Here is another example of the kind of insights we can expect from empirical, sociobiological studies about readers and the act of reading. In this case, the investigators selected 257 female psychology students at a Midwestern American university to respond to four excerpts culled from four British romantic novels, two of which described "proper heroes" while the other two described "dark heroes":

> The results of this experiment support our hypothesis that the dark hero and proper hero in British Romantic literature in the late eighteenth and nineteenth centuries respectively represent cad and dad mating strategies. Modern day college women appear to easily identify the male archetypes represented by these characters nearly two centuries after their creation. . . . Women preferred proper hero "dads," who may be more likely to provide reliable support, for long-term relationships. However, the shorter the relationship in question, the more likely women were to choose dark hero "cads."[100]

The circular logic deployed in this study is stunning, because it does not take the epigenetic effects of social learning into account at all. Its conclusion simply repeats the core premise of the adapted mind without giving due consideration to these students' likely exposure to romantic literary motifs of one kind or another earlier in life, be it during school or via the mass media. The authors may argue that their study provides empirical evidence for EP's theory of mind, yet the only evidence I see is that the entire study was engineered to make that claim seem plausible in the first place. There is just too profound a gap between EP's singular focus on human metaculture, on the one hand, and the multitude of cultural-specific analyses of distinct modes of storytelling informed by a broad variety of social, historical, political, philosophical, and media-theoretical perspectives, on the other. Given the infinite analytical space that separates the two paradigms, one wonders, indeed, "Who Cares about the Evolution of Stories?" as one perceptive scientist put it during the second debate on literary Darwinism in 2012.[101] To answer this question in a typical neo-Darwinian fashion—"we should care because the genes hold culture on a leash, and because the adapted mind predetermines the way (most) readers (are likely to) respond to literary texts"—not only ignores the structural autonomy of cultural evolution as outlined in chapter 3, but also underestimates the unruly potential of meaning that is encoded in linguistic structures and gets activated during the process of reading.

A similar problem haunts empirical studies that do not involve statistical analyses of actual readers' responses to specific texts, but focus instead solely on formal characteristics of particular literary genres. A good example is Jonathan Gottschall's survey of 1,440 folktales from forty-eight different cultures. According to Gottschall, his study provides evidence "that many aspects of gender previously considered products of arbitrary social conditioning are strongly influenced by biology and encountered across human cultures," while, at the same time, it also "testifies to human flexibility and the importance of physical and social environments in influencing the development of individuals and societies."[102] As this wording suggests, Gottschall claims to promote a balanced, middle-path approach allegedly situated between social constructivism on one side and biological determinism on the other. Leaving aside whether this is actually the case, the problem I have with studies like these is their disinterest in specifying the kind of insight literary texts can provide about the nature/nurture problem apart from what we already know—or think we know—from evolutionary biology. Doing so would require a far more detailed analysis of *how exactly* folktales appropriate and *represent* universals themes like love, strife, and death. Gottschall has no interest in doing that.[103] Without that kind of analysis, however, and given the interpretative leverage necessary to draw meaningful conclusions from statistical data, it stands to reason that literary Darwinists will easily interpret this data to confirm, once again, the sociobiological story about cultural evolution and gender relations.

It begs repeating that literary Darwinism is not as revolutionary as it claims to be. A similar approach with similar results dates back to Vladimir Propp's morphological formalism developed in the 1920s and Northrop Frye's archetypal structuralism from the 1950s—neither of which have received much attention in evolutionary aesthetics so far, unfortunately. Although I appreciate evolutionary critics' efforts to flesh out the implied reader using empirical methods, results like those mentioned above will do little to improve the dismal reputation of literary Darwinism in the wider field of literary studies. Carroll and his colleagues, I fear, have missed a great opportunity to forge a meaningful connection between evolutionary aesthetics and the minor history of twentieth-century literary theories interested in embodied readers and the act of reading, such as phenomenology, hermeneutics, and *Rezeptionsästhetik* (the aesthetics of reception).

Hans Robert Jauss is a good representative of this tradition, because he already introduced a theory of "literary evolution" in the late 1960s that emphasized above all the "*socially formative* function" of literature to support human well-being—a claim that seems compatible with the basic premise of today's evolutionary aesthetics. Strongly influenced by Gadamer's model of historical hermeneutics, Jauss was a leading member of the so-called Constance School in Germany that began to flesh out the reader and the reading process in the 1970s and 1980s. He argued that readers' mental aptitude, personal background, and literary expectations profoundly influence the historical meaning of texts. Jauss, and *Rezeptionsästhetik* in general, emphasized readers as active agents rather than passive recipients of textual meaning. The sense of literature, according to Jauss, consists in "the successive unfolding of the potential for meaning that is embedded in a work and actualized in the stages of its historical reception as it discloses itself to understanding judgment."[104] The Kantian legacy in Jauss's theory is unmistakable. Yet Jauss also tried to historicize Kant's transcendentalism by emphasizing the importance of different "interpretative communities" that exist side by side and must determine if, and how, a particular text still engages and makes sense to its (group of) readers.[105] This multiperspectivalism is essential to bioaesthetics as well.

Unlike Jauss, his colleague Wolfgang Iser focused more on the phenomenological experiences and cognitive processes that characterize every reader's individual engagement with the text. His central argument was that each text exhibits numerous blanks and gaps both on the formal level (e.g., temporal or perspectival shifts, changes in narration) and on the level of content (e.g., omissions and allusions, incoherencies and contradictions) that each reader must fill in in order to complete the text and gather its meaning. "The text itself," Iser claimed, "simply offers 'schematized aspects' through which the aesthetic object of the work can be produced."[106] An important conceptual distinction Iser introduced to clarify this idea was that between the written "text" and the literary "work." Whereas the text remains the same regardless of whether it is being read, the literary work gains different contours and acquires new meanings during the act of reading.

Literary Darwinists, unfortunately, never acknowledged or engaged the aesthetics of reception. Instead, they chose to lump together all strands of literary theory since the 1960s under the single paradigm of "Foucauldian cultural critique." Whatever that means, it fails to distinguish between German *Rezeptionsästhetik,* based on the hermeneutic tradition, and French deconstruction, based on (post-)structuralist theory. The difference between the two paradigms can be exemplified by their opposing views of the relation between "work" and "text." In 1971, Barthes had already introduced this very distinction in an essay titled "From Work to Text." As his title indicates, Barthes emphasized the text, not the work, and his understanding of these terms and how they relate to one another is the exact opposite of Iser's. For Iser, the text is finite and concrete, whereas the work is virtual and abstract. For Barthes, it is precisely the opposite: "the work is a fragment of substance," he claims, whereas "the Text is a methodological field." The work is closed, while the text remains open, "or again, *the Text is experienced only in an activity of production.*"

Iser's sense of the relation between "text" and "work" strikes me as far more intuitive than does that of Barthes. Only a reader of Barthes's remarkable ingenuity would choose to characterize a concrete material object in space—say a written text or a book—as a "constitutive movement," "a process of demonstration," and "an activity of production."[107] These vitalist terms misleadingly imply that texts exist sui generis and are able to operate without any help from the outside—an implication made explicit by other deconstructive critics like Paul de Man and Jacques Derrida.[108] Yet texts are neither autopoietic nor conscious. And they cannot move unless they are being moved. A written text is not an activity or a process of any kind, except, of course, when a reader actively engages with it—that is, when a text is actually being read *by someone*. But that "someone," in Barthes's view, is not an embodied reader at all, but merely an operating function or structural effect of language itself. His theory, and French (post-)structuralism in general, returned again and again to the scene of writing as opposed to the process of reading. Iser's hermeneutic model, by contrast, puts sense making back in the active mind of the reader where it belongs. It is not language that speaks, but embodied readers. It is they who actively produce the literary work and its meaning by filling in the empty spaces within the text.

The aesthetics of reception was relatively short-lived. Phenomenological-based studies of reading more or less disappeared during the 1990s and have proven far less influential than structuralism or deconstruction. Given the humanities' current fascination with anything deemed nonhuman, ahuman, transhuman, or posthuman, this is unlikely to change anytime soon. This lack of interest in embodied readers, however, renders literary Darwinists' disregard of *Rezeptionsästhetik* all the more disappointing. Despite their empiricism, Carroll and his colleagues consider the sociohistorical background of today's readers far less important than that of dead authors. Authors' biographies matter a great deal to them; readers' biographies do not. The meaning of texts, in their mind, does not reflect the norms and values of particular social communities, but the inherited architecture of the "universal people."[109] This dichotomy has prevented literary Darwinism from developing studies about the interpretative bias or genre expectation of empirical readers (or communities of readers) as opposed to ever more studies that confirm their view of "human nature." The former would be innovative and intriguing; the latter is trivial and entirely predictable.

Cognitive Studies

Our discussion of evolutionary aesthetics so far has focused on reductionist approaches inspired by E. O. Wilson's concept of sociobiology and his notion of consilience. The second group of scholars I want to discuss in this final section embraces a far less rigid understanding of evolutionary theory. "Recent cognitive studies," Ellen Spolsky remarked in 2010, "endorse a view of human life within the material world as dynamic and relational rather than as static and hierarchical."[110] A wide-ranging and complex field, the cognitive approach to art and literature has resisted the premature attempt of normativization that ruined literary Darwinism. Neither sociobiology nor consilience plays a central role in cognitive studies, and orthodox EP is just one among many (and more nuanced) versions of evolutionary psychology relevant in the field. Dating back to the 1980s, the rich history of interdisciplinary debates in cognitive studies has since yielded a level of methodological sophistication and research productivity impossible for me to summarize in detail.[111] My goal instead is to introduce the basic methodology and core arguments of the field, including its firm commitment to disciplinary autonomy and a shared resistance to the "imperialistic take-over" attempts that characterize overly reductionist approaches in evolutionary aesthetics.[112] Rather than chasing after Wilson's "unified theory of knowledge," these scholars welcome the disciplinary heterogeneity that informs current research in the humanities. In their view, the methodological incongruity between the two cultures does not and should prevent their cooperation.[113]

The leading scholars of the cognitive approach (like Ellen Spolsky, Alan Richardson, Patrick Colm Hogan, Lisa Zunshine, H. Porter Abbott, Mark Bruhn, Nancy Easterlin, and many others) have developed an open, flexible, and multifaceted methodology that draws not only from the life sciences and evolutionary studies, but also from linguistics and philosophy, including hermeneutics and deconstruction. Their objects of study are varied, too, and range from single concepts (like "compassion" or "the implied author") to whole periods and genres (like British romanticism or realist novels).[114] But whatever their focus, these critics consistently pursue a two-way exchange between science and hermeneutics, between theories and texts, because they believe that cognitive science has as much to learn from literary studies as vice versa.[115] It follows that cognitive studies of art are significantly more interested in how specific artworks challenge or complement, rather than merely exemplify, universal cultural traits or preconceived notions of human nature. As Spolsky explains, cognitive literary theory studies the relation "between the humanly universal and the culturally and individually specific, as coded and recorded in cultural artifacts."[116] This concise statement of method applies to bioaesthetics as well, because it zooms in on the infinite coding and recording processes that characterize cultural evolution. These processes are semiotic as much as genetic in nature, and it makes little sense to quibble about which of these is more important when both are essential. This insight into the epistemological equality of the humanities and the sciences clearly distinguishes the field of cognitive aesthetics from biologism. Unlike the latter, the former is, above all, adaptable, in the sense that cognitive aesthetics remain sensitive to the specific environment and sociohistorical context of art and culture.[117] Although the members of this group remain indebted to Darwinian evolutionary theory and cognitive theories of mind, most of them are critical of orthodox EP and its rigid model of mental architecture.

Patrick Colm Hogan's work is exemplary in this regard, because it showcases the potential conflict between these two tendencies. On the one hand, Hogan's view of "literature as a body of data" clearly echoes the research paradigm espoused by Carroll and colleagues in their recent empirical survey of nineteenth- and early twentieth-century British novels discussed above. Hogan's book, too, is based on the premise that "the data [literature] provides can and should be used to complement empirical studies in the sciences."[118] The explicit goal of his approach is to develop "a theory of literary universals [that] would describe a *repertoire of techniques* available to authors and a range of *nontechnical correlations* derived from broad statistical patterns."[119] On the other hand, however, Hogan is fairly critical of literary Darwinism and EP, and his research includes close readings of individual texts. This approach reflects Hogan's premise that the art of interpretation is an essential part of any and all analysis, in literature as much as in the sciences. Literature, Hogan explains, yields a unique kind of data, the kind that "present[s] us with detailed interpretative contexts for comprehending the mechanical explanations articulated by the scientists." It follows that literary texts, in Hogan's view, not only depict or represent generic human emotions, but also affect and shape our emotions in at least three ways: first, they evoke readers' affective response to the text; second, they enable readers to reflect and elaborate on "our prototypes for persons, events, actions, and conditions"; and finally, they influence readers' social behavior.[120]

On the methodological level, Hogan thus grants literature—and literary studies—the power to influence the affective range of human emotions. He posits a two-way street between literature and science, which clearly separates his cognitive approach from literary Darwinism. Instead of using literary data exclusively to demonstrate the universal nature of human emotions, Hogan wants to explore "how literature and literary studies themselves . . . may alter our spontaneous emotional responses to the world and to ourselves, and our imaginative elaboration of those responses."[121] On the interpretative level, however, Hogan does not always live up to this goal. His readings of literary texts frequently and unduly stress the degree to which fictional characters' emotions and actions end up confirming "just what we would expect" given our scientific knowledge about emotions. Hogan's interpretation of Shakespeare's *Romeo and Juliet*, for example, proceeds along the premise that the heroes' emotional reaction is archetypical for people in love and operates within the empirically established parameters of average—or normal—human behavior. *Romeo and Juliet*, he insists, "is more or less [a] universal legend of love."[122] This premise prompts Hogan to interpret any discrepancy he detects between scientific–empirical studies of human emotions and their dramatic representation on stage as a sign of the aesthetic *weakness* of the play. Confronted with parts of Shakespeare's drama that are difficult to reconcile with the universal patterns Hogan expects to find, he readily faults the artwork, not the theory, for its alleged failure to make sense to the reader.

Hogan's interpretation of the famous balcony scene is paradigmatic in this regard. According to Hogan, Romeo's confessions of love in this passage are so bizarre as to become incomprehensible to the common reader: "Despite its renown, most of what Romeo says in this scene is adolescent nonsense (e.g., that her eyes have gone to heaven to twinkle in the place of stars—a bizarre image)." The obvious problem with this reading is that if literary scholars were to follow Hogan's advice and dismiss Romeo's "proneness to muddled poetic conceits" as adolescent gibberish unworthy of further investigation, there would remain nothing to be learned from this fictional character's speech and behavior—except, of course, what we presumably already know from scientific studies of how most people in love actually behave in real life.[123] If folk psychology and traditional theory of mind provide the sole criteria by which to determine whether literary texts make sense to "the reader," then the kind of aesthetic dyslexia Hogan occasionally displays in his readings will likely become the norm. Yet why study the emotional life of literary characters at all if their behavior and expressions cannot, or must not, disagree with our empirical understanding of human affect?

Lisa Zunshine addresses this question head-on in her 2006 monograph titled *Why We Read Fiction*. According to Zunshine, the main reason we enjoy reading is precisely that literature commonly does *not* confirm what we already know, but, on the contrary, challenges our habits of normative behavior and cognitive patterns: "The moment cognitive scientists succeed in isolating yet another plausible regularity," Zunshine argues, "we can start looking for the ways in which fictional narratives . . . have been burrowing into and working around that regularity; testing and reconfiguring its limits."[124] Methodologically speaking, Zunshine, like Hogan, grants fictional texts significantly more autonomy than literary Darwinism. In her 2010 anthology, Zunshine introduced the term "cognitive cultural studies" to designate the new research paradigm she envisions. Based on the "convergence between early cultural studies and recent cognitive theory," the explicit goal of cognitive cultural studies is to "open new avenues for investigating the role of universally shared features of human cognition in historically specific forms of cultural production."[125] Note Zunshine's use of "convergence" rather than "consilience" in the quote above. Her main historical reference for defining "cognitive cultural studies" is the Marxist Raymond Williams, not the sociobiologist E. O. Wilson. Williams's notable interest in Darwinian theory demonstrates, in Zunshine's view, that "a cognitive–evolutionary perspective commits cultural critics to a more rigorous historicizing than a perspective that ignores human evolved cognition."[126]

Given her methodological commitment to intellectual pluralism, however, it is surprising to find that Zunshine's readings are mostly informed by key concepts drawn from EP's theory of mind and Steven Pinker's work in particular, even though she aptly notes Pinker's inadequate understanding of literary modernism.[127] Still, there is little reference to Varela and Thompson's influential notion of the embodied mind, nor to Lakoff and Johnson's theory of enactment and cognitive metaphors, nor to any other socioempirical study of human behavior not based on orthodox neo-Darwinian theory. In stark contrast to German reception theory, which emphasizes the plasticity of readers' responses to literary texts regardless of period and genre, Zunshine's study, similar to literary Darwinism, tends to privilege a particular group of literary texts—namely, those that "engage clusters of cognitive adaptations associated with our ToM [theory of mind] and metarepresentational ability in a particularly focused way."[128] These texts are frequently novels, and nineteenth-century realist novels to boot. Poetry, short stories, or minimalist literature of various kinds (e.g., fables, anecdotes, parables) are underrepresented genres in Zunshine's repertoire. Her selection criteria end up being too prescriptive, in my view, and will likely bias literary interpretations to confirm rather than challenge the intellectual intuitivism championed by EP and sociobiology.

Despite their innovative interdisciplinary approach, the studies by Hogan and Zunshine demonstrate how difficult it is for cognitive cultural studies to match its methodological commitment to pluralism with textual interpretations that follow suit. The major challenge for bioaesthetics, too, will be to come to terms with the conflicting forces of cultural variation and genetic constraint encoded within cultural artifacts. The best way to do this, in my view, is to focus on the increasing plasticity of the brain. Ellen Spolsky, too, is a strong proponent of brain plasticity: "Plasticity and flexibility are important characteristics of healthy brains because many early-built connections will not serve for a lifetime," she aptly points out. "The dynamic that allows ongoing realignment is built into the system; it is the logical outcome of evolutionary development."[129] This means that every human brain continues to adapt throughout the entire lifetime of the phenotype and not just during the critical phase of infancy and early childhood. This warrants the conclusion that cultural evolution increases the brain's inherent evolvability—that is, its ability to increase efficiency by raising the level of plasticity built into the system at the neuronal level.

If this assumption is correct—and the next chapter will make the case—what are the consequences for the study of art and culture? For one, the brain's increased plasticity due to cultural evolution suggests that phenotypes are far less restricted by their inherited mental architecture than neo-Darwinian theory, including sociobiology and EP, stipulates. Spolsky makes the same claim I have made throughout this book, namely, that the cognitive effects caused by cultural evolution "are culturally heritable: they can and have changed the cultural environment in ways that give later generations a head start." She also emphasizes that artists are particularly sensitive to cultural evolution and technological change, which explains why their works often challenge our habits of thinking, our naturalized categorizations, and the way we "normally" see things. Hence the best works of art are often those that defy common sense and folk psychology rather than confirm it. We might say that artists activate the built-in plasticity of our sensorium—that is, the fact that "different modalities of representation . . . are able to teach different things by recruiting different modalities of sensory understanding." Spolsky's understanding of the cognitive function of art is representative of bioaesthetics at large. Its premise should be that works of art do not offer solutions to scientific or metaphysical problems; instead, they display "misalignments and prompt the search for solutions."[130] Spolsky and Richardson demonstrate that the best work currently done in the field of cognitive cultural studies is informed by recent discoveries about the brain's neuronal plasticity. The rise of neuroaesthetics in particular challenges key concepts and methodological assumptions of literary Darwinism and neo-Darwinian aesthetics, as I will demonstrate in the next chapter.

5 NEUROAESTHETICS

The brain is the screen.

GILLES DELEUZE

How to "Read" a Brain Scan

On November 11, 2007, the *New York Times* ran a full-page article in its Op-Ed section titled "This Is Your Brain on Politics."[1] Cowritten by three neuroscientists, one public policy analyst, and three neuromarketing experts, the article summarized the test results of using fMRIs on twenty undecided voters who had to answer questions about their preferences for the upcoming 2008 presidential election. The results were hardly revolutionary: "Voters sense both peril and promise in party brands" was one of the findings in bold print. "Emotions about Hillary Clinton are mixed," "The gender gap may be closing," and "Mitt Romney shows potential" were some of the others.

The reason why this story made it onto a major page of the *New York Times* instead of being relegated to the horoscope section of your local newspaper was the scientific evidence presented to back up the claims. This evidence consisted of numerous color pictures that showed the activation levels in human brains when confronted with the words "Democrat," "Republican," and other simple statements or queries. Each of these brain scans highlighted particular sections of the cerebral cortex in orange and yellow, thereby indicating the higher percentage of oxygen in the blood flow of these activated regions due to higher brain activity in those particular areas. Scientists interpret this correlation to mean that activated nerve cells require more oxygen than those in a state of rest. This premise informs the current use of fMRI to measure human brain activity. The method is based on the fact that oxygen-rich hemoglobin has different magnetic characteristics than oxygen-poorer hemoglobin, which allows the scanner to localize and measure the magnetic difference between active and passive areas of the brain. The resulting data is transformed into a digital image of the brain according to preprogrammed computational codes that provide color, perspective, shading, et cetera. The result is a "picture" of your brain while engaging a particular task or focusing attention on specific words, images, or ideas.

Politics, of course, is not the only subject increasingly studied with the help of neuroimaging technology. In October 2007, for example, the journal *Scientific American Mind* printed an article titled "Searching for God in the Brain." The author stated that "researchers are unearthing the roots of religious feeling in the neural commotion that accompanies the spiritual epiphanies of nuns, Buddhists and other people of faith."[2] In 2004, the neuroscientist Helen E. Fisher claimed that we "are coming to some understanding of the drive to love," a passion that "emanates from the motor of the mind, the caudate nucleus; and it is fueled by at least one of nature's most powerful stimulants, dopamine."[3] Over the last two decades, there has been a plethora of similar proclamations by scientists about the neurological nature of our most intimate dispositions, feelings, and beliefs. At stake in this inquiry, Jean-Pierre Changeux argues, is nothing less than *The Physiology of Truth*.[4] The central claim is that our feeling of selfhood and subjectivity can objectively be linked to—and possibly deduced from—the electrochemical activity in our brains. "You are your brain" is how the neuroscientist Michael Gazzaniga summarizes the premise of many of his colleagues and the field at large.[5] Or, as Francis Crick put it in his *Astonishing Hypothesis*, "You are nothing but a pack of neurons."[6]

Neuroimagery remains popular both within and beyond the academy, yet the original euphoria has given way to a greater awareness of the limited insight these "pictures of the brain" provide into the human mind. As Kenneth W. Goodman, a bioethicist at the University of Miami, laconically states, "We don't really know what parts of your brain lighting up really mean."[7] Due to technological constraints, for example, researchers have to choose between accurately measuring either the temporal or the spatial dimension of neural processes, since to date there exists no single method able to provide precise data for both time and location.[8] Another major problem is that these "scans typically are normalized to a standard stereotactic space" in order to average out distortions caused by spontaneous and irregularly occurring impulses unrelated to the measured activity. This common practice, however, ignores the fact that subjects' brains often differ "both in relative locations and sizes of specific regions" being measured.[9] Moreover, since the brain is always doing something, scientists have to subtract the normal blood flow "at rest" (i.e., the average brain activation levels during control tasks) in order to measure the exact increase of blood flow activated by the particular task in question. Given the difficulty of quantifying this state of "rest" for any particular brain, this adjustment typically relies on yet another statistical average mixed into the picture. In fact, the final picture presented to the public is usually a compilation of dozens of individual scans taken from various groups of human individuals. This also applies to the *New York Times* essay mentioned above. Even though the article talks about "your" brain on politics, one of the captions cautions the reader that "the images do not represent individual brains, but rather reflect the combined data gathered from several—in some cases all—subjects."

The crucial point is that despite their common pictorial form, fMRI scans are not iconic, because their material referent (i.e., cranial blood flow) remains invisible to the naked eye. The neurosciences do not "take" pictures of the brain; they "make" pictures of the brain. For these pictures have no referent other than the numerical data set that informs them, and there is no scientific reason to represent that data in the form of colorful images as opposed to, say, a column, a chart, a graph, or a written text. In fact, a simple graph or numerical chart would represent the underlying data more accurately than the colorful pictures we are accustomed to seeing. The main difference, of course, is that a string of numbers carries neither the emotional appeal nor the implicit truth claim we intuitively attribute to pictorial representations of the human brain. Digital imaging technologies, in other words, effectively serve as evidence-producing machines: they present colorful pictures of hitherto insufficiently understood processes and cognitive functions *as if* these pictures had explanatory powers—which, alas, they do not possess. Like any and all forms of representation, neuroimages require interpretation by experts to become meaningful, while, at the same time, their ostensible iconicity and self-evidence serve to disavow this requirement.[10]

In media-historical terms, we might say that brain-image technology has managed to usurp the erstwhile reality effect of traditional photography—the fact that "this has been," as Roland Barthes famously put it.[11] This is what new media always do: they remediate old media.[12]

Whenever a new medium of representation enters the historical stage, it quickly assimilates and usurps the ideological power of the older media it seeks to replace. At the same time, however, new media must also accommodate the metaphorical and conceptual terminology they inherit from their predecessors. In 1844, William Henry Fox Talbot idealized photography as "the pencil of nature" much like neuroscientists today idealize fMRIs as "pictures of the brain." This constant give and take between old and new media also explains the continuing influence of film theory and aesthetics on contemporary GUI and web design, evident in our common reference to computer "windows," "frames," "screenshots," et cetera.[13]

I want to suggest that twenty-first-century digital neuroimagery appropriates the representational power of traditional photography in much the same way that nineteenth-century photography appropriated the representational power of traditional painting. If early photography sought to legitimize itself with reference to writing and the art of drawing, today's brain imagery seeks to legitimate itself with reference to traditional photography and other forms of pictorial representation. And just like traditional painting was unable to compete with the heightened reality effect of photographic realism, traditional photography today cannot compete with the heightened sense of scientific objectivity captured in a brain scan.

My point, of course, is not to deny the benefits of fMRI technology for medical research and patient treatment. I merely seek to elucidate the rhetorical power of these pictures, which stems largely from their implicit promise to render visible the invisible—that is, to depict a hidden reality we cannot normally perceive. Yet precisely for this reason, it is all the more important to remember the phantasmagoric nature of these images. Contrary to appearance, the history of media—like the history of science—is not progressive: it has no origin and no telos.[14] New media are not necessarily superior to or better than the old media they replace. Instead, different media capture different aspects and various degrees of verisimilitude between reality and its representation. In epistemological terms, these differences are irreducible and nonquantifiable. They be cannot be ranked or hierarchized according to universal, objective criteria of better or worse, "true" or "false."[15] Unlike biologism, bioaesthetics must have no illusions about the fact that no symbolic system or medium of representation will ever match the infinite complexity of the real world. The best we can hope for is to approximate this complexity one way or another. Yet none of them, nor all of them combined, will ever add up to the whole picture.

In the previous two chapters, I have demonstrated the debilitating effects of biologic reductionism if applied to the study of literary texts and cultural evolution at large. This chapter makes a similar argument with reference to some versions of neuroaesthetics, while, at the same time, it highlights the considerable accomplishment and intellectual promise of others. Although neuroaesthetics might be considered part of evolutionary aesthetics in the larger sense, it has clearly emerged as the single most prominent field under the current hegemony of the life sciences. Several major anthologies and monographs on the relation of art and brain have been published in the last few years alone, not to mention the plethora of other books, scientific articles and critical reports in academic journals and the mass media since the 1990s.[16] A major reason for the popularity of neuroaesthetics is that recent advances in bioengineering have made it possible to study people's brains while they engage in aesthetic behavior of various kinds such as reading or writing, watching or drawing images, hearing or playing music, et cetera. The increasing number of academic institutions and laboratories all over the world dedicated to the neuroscientific study of art and artistic behavior indicate that neuroaesthetics is about to establish itself as a genuine discipline in its own right.[17] Yet precisely because of its enormous academic and popular success over a relatively short period of time, the new discipline still lacks a clear sense of how the different approaches and distinct objectives collectively summarized under the term "neuroaesthetics" relate to one another. Hence the overall goal of this chapter is to sharpen the overall contours and various strands within the discipline in order to differentiate their major concepts and distinguish their methodologies.

I begin with a historical survey of major breakthroughs in human brain research since 1800 in order to introduce the central concepts and remaining problems in contemporary neuroscience and the study of consciousness. After that, I briefly discuss two distinct theories of human consciousness—namely, the functionalist and the phenomenological point of view. Although not directly related to the study of art and culture, these two sections are important, in part because neuroaesthetics cannot be conceived outside of its larger historical–theoretical context, and in part because cognitive theory plays a central role in bioaesthetics overall.

The two sections after that zoom in on the two main approaches within contemporary neuroaesthetics—namely, historical-oriented approaches focused mainly on close readings of individual artworks, on the one hand, and scientific studies largely based on empirical–statistical data, on the other. What separates the two, I argue, is a methodological shift from reductionism to emergentism and the concomitant analytical shift from brain modularity to brain plasticity. At the same time, I hasten to add that reductionism and emergentism, much like modularity and plasticity, must not be defined as mutually exclusive processes or concepts. They should be conceived instead as complementary aspects of how the brain works to produce knowledge. Put differently, I want to emphasize that each of these two main approaches includes a variety of methods and perspectives, some of which might actually have more in common with specific strands in the other group than with members of their own.[18]

The field of neuroarthistory provides a case in point. John Onians, who introduced the term neuroarthistory in 2008, is an art historian, and most scholars who have since adopted the term and contributed to the movement are likewise humanists, not scientists.[19] They formulate a neurological approach to art and art history that is neither data-driven nor based on brain imagery, but emphasizes instead the plasticity of the human brain as a means to elucidate the prevalence of particular styles and motifs that characterize the work of individual artists. Despite this historically minded and nonempirical approach, however, neuroarthistory features the same kind of revolutionary rhetoric and totalizing perspective that we already encountered among literary Darwinists and that I deem incompatible with bioaesthetics. "The particular strength of neuroarthistory," Onians claims, "is its ability to reconstruct the unconscious intellectual formation of the makers, users and viewers of art."[20] In fact, the German art historian Karl Clausberg had already made a similar statement in his monograph *Neuronale Kunstgeschichte (Neuronal Art History)* from 1999: "Indeed, very many pictures that seem to merely depict external reality or bizarre imaginary worlds unconsciously represent the central structures of cerebral organization of their producers."[21] Both Onians and Clausberg argue that paintings reflect the particular neurological structure of their creators' brains, and that not only artists of all traits (dead or alive), but anybody interested in or engaged with the phenomenon of art is essentially reducible to his or her brain.

Equally disappointing, in my view, is that contrary to Onians's revolutionary claims, his approach is fairly traditional, because it relies mostly on biographical information about individual artists and critics, a common practice in art history since the nineteenth century. Thus Onians, too, begins his historical survey with a detailed account of Aristotle's life and poetics. Aristotle, he argues, "was the first person to suggest that the making of, and the response to, art might have a biological, even a neurobiological basis"—despite the fact that Aristotle never dissected a human brain, was unfamiliar with the nerves, and falsely believed that the heart (and not the brain) was the seat of sensations and the intellect, as Onians himself points out. The major function of the brain, according to Aristotle, was to cool the organism. Still, Aristotle's intellect, Onians claims, was due to "his distinctive upbringing" along with the fact that he was "living with his medical practitioner father" in "a variety of environments" that brought him in "much closer contact with an even greater variety of wildlife, both flora and fauna, than the typical urban Athenian."[22] It was this particular sociocultural environment that shaped the particularity of "Aristotle's neural apparatus" and determined his thinking. I have no doubt that this is true. Yet Onians's singular focus on the neurological specificity of people's brains adds little else to what we already know from traditional sociohistorical approaches about the influence of biographical and environmental factors on individual artists and their artistic creations—except for identifying the brain as the specific locale where these factors leave their permanent mark. Apart from its neurological premise and scientific nomenclature, there is little to distinguish neuroarthistory from the established discipline of art history.

My crucial point is that these terminological distinctions—including my own—primarily serve a heuristic purpose and are far from absolute. As our focus slides up and down the analytical scale from microbiological processes to individual artworks to cultural evolution writ large, it brings into view both similarities and differences within and between the scientific–empirical and the hermeneutic–historical approach to art and aesthetics. Hence the different versions of neuroaesthetics I discern in this chapter can be seen as contradictory or complementary depending on the particular conceptual framework used to compare them. This methodological flexibility, as well as the epistemological ambiguities that arise from it, are constitutive of bioaesthetics and constitute a principle to be cherished, not a problem to be solved.

The last section of this chapter is dedicated to the work of Semir Zeki, an accomplished and internationally renowned neuroscientist who heads the one and only Institute for Neuroesthetics, at University College London. There can be little doubt that Zeki's scientific research has fundamentally shaped the field of neuroaesthetics at large. His studies of the human brain and visual processing over the past thirty years have become standard references in the medical community, and his book *Inner Vision* from 1999 is widely considered one of the most successful attempts to bridge the two cultures. While I generally agree with this assessment, I also argue that Zeki skillfully fuses scientific materialism with speculative idealism in a way reminiscent of what Louis Althusser, back in 1968, called the "spontaneous philosophy of the scientists" (see chapter 2).

The Cerebral Subject

According to Semir Zeki, the main objective of *neuroesthetics* (note the British spelling!) is to outline "a theory of aesthetics that is biologically based." This premise clearly resonates with the notion of bioaesthetics I developed throughout this book. Yet Zeki's definition somewhat understates the strong neurological focus on the brain that distinguishes his view of the discipline from both the broader anthropological framework of evolutionary aesthetics and from bioaesthetics overall. It would be more apt to say that neuroesthetics seeks to develop an aesthetic theory *based on the brain*. The brain, Zeki emphasizes again and again, is the central organ in control of body, mind, and our overall appreciation of art. Zeki believes that artists, like all human beings, are essentially reducible to their brains. Painters, for example, work and rework "a painting until it achieves a desirable effect, until it pleases them, which is the same thing as saying until it pleases their brain."[23]

Zeki's work proves—if proof were necessary—that the twenty-first century is the century of the brain.[24] Critics have coined the terms "cerebral subject" and "neural subject" to describe the crucial premise in much of contemporary cognitive science and neurophysiology that the human being is essentially reducible to his or her brain.[25] The long history of this cerebral or neural subject, however, dates back to the work of the Viennese anatomist Franz Joseph Gall. In the last years of the eighteenth century, Gall started giving public lectures on his highly influential theory of cranioscopy, which was later renamed phrenology by Gall's erstwhile disciple and collaborator Johann Spurzheim. Similar to Johann Kaspar Lavater's earlier writings on physiognomy—that is, the attempt to determine a person's character based on his or her facial features—Gall's theory stated that the external shape of a human skull provides information about the moral character and innate psychological disposition of its owner. Although many aspects of this imprint theory were discredited and ridiculed even during Gall's own lifetime, its basic premise remained influential throughout the nineteenth century and even received qualified support in Darwin's *The Expression of the Emotions of Men and Animals* from 1872. Phrenology also proved particularly important for the rise of criminology throughout Europe at the time, playing a key role in Cesare Lombroso's pseudo-scientific studies of *L'uomo delinquente (Criminal Man)* published in 1875 and in Hanns Gross's *Criminalpsychologie (Criminal Psychology)* from 1898.

Gall's basic idea about the modular structure of the brain still remains relevant today. Gall was the first to speculate that our mental abilities emerge from different spatial locations inside the brain, each of which, he thought, is specifically "designed" for particular tasks like sensation or cognition. Gall's crude version of cerebral localization has since been falsified, and it is commonly accepted in neuroscience today that most cortical areas contribute to a variety of cognitive functions via a process called neural redeployment. Nonetheless, Gall is still considered an important early theorist of anatomical brain modularity—that is, the idea that sensation and cognition are regulated by, and strongly correlated with, the activation of one or several discrete cortical areas.[26] Brain modularity continues to play a major role in cognitive neuroscience today, as we saw in our discussion of EP and its massive modularity theory about the adapted mind (see chapter 3).

Most neuroscientists, unfortunately, rarely reflect on the historical roots of their discipline dating back to the nineteenth century.[27] Yet the fact remains that despite the many modifications since then—including the early realization that the skull's bone structure is totally irrelevant in terms of brain function, or the discovery of neuron cells and the formulation of the single neuron doctrine around 1900, or the increased focus since the 1970s on neuronal binding to study how synapses operate and give rise to large neuronal maps—Gall's localization theory still informs the basic conceptual framework for today's understanding of the human brain. Today's neurosciences, critics lament, are rich in data but poor in theory: "The list of articulated models is extensive, yet theoretical understanding remains piecemeal."[28] The striking discrepancy between neuroscience's revolutionary rhetoric based on new technologies to visualize scientific data, on the one hand, and the far less compelling methodological and theoretical foundations of this rhetoric, on the other, remains a major point of criticism within the discipline and beyond.[29]

The methodological problems haunting the discipline, in other words, far exceed its reliance on digital technology and the lack of measuring accuracy I mentioned above. For one, there are serious limitations about the kind of information scientists can glean from studying impaired brains as opposed to healthy ones. Yet the vast majority of clinical studies throughout the twentieth century were actually based on patients with various kinds of brain injuries due to war, accidents, or crimes, and for obvious reasons: it would have been unethical to conduct invasive medical experiments on healthy brains (human brains, that is; we do it to animals all the time). For example, there is ample empirical evidence that aphasia—the loss of the ability to use and understand language correctly—is associated with injuries to two specific areas of the cerebral cortex: the inferior frontal gyrus called Broca's area (which governs language production), and the superior temporal gyrus called Wernicke's area (which governs the interpretation of language). To this day, however, researchers cannot positively conclude that the affected areas are solely responsible for the observed linguistic malfunctions, nor do they know how Broca's and Wernicke's areas interact with other regions in the brain that may also contribute to the correct use of language.

A core problem behind many of these uncertainties is the limited knowledge about the "mid-level" organizational structure of the brain and the ways smaller or larger clusters of neurons interact with one another. Scientists agree that our knowledge of this mid-level lags far behind our understanding of larger areas (e.g., the amygdala, the hippocampus, or the thalamus) and that of smaller brain units (e.g., single neurons). In a manifesto from 2006 cowritten by eleven leading neuroscientists, the authors provide a well-balanced assessment of the problems and promises of neuroscience. Their reservations are still valid today and worth quoting at length:

> Without doubt, we do know a lot more today about the brain than we did ten years ago. However, there still exists a major epistemological gap between the upper and the lower organizational level of the brain. We know frighteningly little about what happens at the mid-level—the level that comprises smaller and larger networks of neurons—that forms the basis of those processes taking place at the highest level. Likewise, we can only guess about which codes are used by individual neurons to communicate (they probably use several of such codes simultaneously). It is completely unknown what happens when several million or even billions of neurons "talk" to one another. The rules according to which the brain works; how it succeeds to represent the world such that immediate perceptions and earlier experience melt together; how the inner processes are configured as "its" dealings and how it

plans future actions—none of this we have even begun to understand. Even worse, it is not clear how we could possibly research any of this using today's means and equipment. In this regard, we are, so to speak, still at the level of hunters and gatherers. The description of activation centers via PET or fMRI and the correlation of these areas with particular functions or activities does not help much. The fact that all of this takes place at a particular spot in the brain does not constitute a real explanation of these phenomena. For these methods say nothing about "how" all this works. They only measure very indirectly a slightly higher use of energy at a certain spot within a heap of hundreds of thousands of neurons. This is the same as if one were to try to understand the functioning of a computer by measuring its energy use while it performs various tasks.[30]

Confronted with this impressive list of what we *do not yet know,* one may start to wonder what, exactly, we *do know* about how the brain works. In the following, I shall focus on our sense of vision and the visual cortex, which is one of the most studied parts of the brain and highly relevant for our discussion of neuroaesthetics. To begin with, clinical trials have shown that each neuron reacts only to stimuli of a particular kind (such as movement, intensity, orientation, or frequency) and only if these stimuli cross an exceedingly small area called the receptive field. Exposed to the wrong stimuli or stimuli outside their receptive field, vision neurons show little or no response. The question of how exactly the "right" picture of the world is put together from these evidently disparate bits of neuronal information remains largely unanswered. We do know, however, that the necessary interpretation of data does *not* just take place at the very end of the line, so to speak, in the neocortex. Contrary to centuries-old philosophical speculation, there is no determining agency able to put all the pieces together and bestow the final meaning on data: "We must stop thinking of the brain as if it had such a single functional summit or central point," Daniel Dennett insists. There is no "Central Meaner" or "Cartesian Theater" that authorizes meaning. Rather, Dennett argues, "at any point in time, there are multiple 'drafts' or narrative fragments at various stages of editing in various places in the brain."[31]

The crucial point is that the external world is not mimetically mirrored in the perceptual–cognitive system of living organisms. Neuronal maps in the human brain certainly correlate to external objects, but there is no one-to-one equation between a specific neuronal state and the object represented by that state, because the same object can be linked to different states, and vice versa: the same state can be linked to different objects. Neuronal representation is highly dynamic, associative, and adaptive. This plasticity of the brain's neuronal network was originally discovered by the neurologist Stephen Kuffler in the 1950s during his experiments on retinal ganglion cells (i.e., the first layer of neural tissue that rims the eyeball). Kuffler found that a large spot of light shone on the cells created less of a response than just a small spot of light. The reason, he surmised, was that the receptive field of each cell was organized by a center–surround structure, meaning that cells do not react uniformly to a stimulus within their receptive field, but also take into account the stimuli differential among different areas inside the field. The center–surround structure of single neurons means that visual perception is an active construction, not a passive recording, of external stimuli. Hence neuroscientists speak of image *processing,* not image *transmission.*[32]

A major consequence of this research was that psychologists today have largely abandoned the traditionally held distinction between sensation and perception. According to this earlier model, our senses record the incoming stimuli as unorganized data, which is subsequently processed and interpreted by the cognitive system of the brain. Today, however, we know that the imaging process is spread out over the entire visual system, which begins with our sense organs and only ends in the higher cortex. This means that sensation *is* perception, because "processing sites are also perceptual sites," as Semir Zeki points out.[33] Contrary to Descartes's famous dictum—"it is the mind that sees, not the eyes"—we should say that it is *both* the mind *and* the eyes that see. In a seminal paper from 1968 titled "What the Frog's Eye Tells the Frog's Brain," a group of scientists including Humberto Maturana concluded after a series of experiments that "the eye speaks to the brain in a language already highly organized and interpreted, instead of transmitting some more or less accurate copy of the distribution of light on the receptors." Retinal cells provide the brain with more than just a bunch of disparate data. Instead, they actually map out a congruent space according to the specific operation they are programmed to record. It follows that "every point is seen in definite contexts. The character of these contexts, genetically built in, is the physiological synthetic a priori."[34] The implicit reference to Kant ("synthetic a priori") is hardly accidental in my view, because neuroscientific research essentially confirms Kant's transcendental–constructivist view of how we perceive the world. As Maturana puts it, "We enact the world we live in."[35]

The constructivist nature of human perception and cognition has important implications for brain plasticity. One of the positive effects of fMRI imagery, in fact, has been to modify neuroscience's traditional emphasis on the modular structure of the cerebral cortex. Today, the focus instead lies on the brain's ability to create new synaptic pathways in response to environmental changes in postnatal life. Although the brain is certainly not a tabula rasa, its neuronal network is largely formed in response to external stimuli after birth. This insight dates back to the groundbreaking experiments of David Hubel and Torsten Wiesel during the 1960s. Based on animal research, Hubel and Wiesel proved that the brain grows in response to external stimulation during the "critical" period of neural development right after birth. In his Nobel Prize acceptance speech from 1981, Wiesel explained "that a few days of monocular closure had caused clear physiological and anatomical changes in the striate cortex" of newborn cats.[36] Citing these and similar experiments as well as his own research, Zeki concludes that "there is a critical time during which adequate stimulation is mandatory if the animal is to be able to see at all." Even more important is the fact that neuronal plasticity extends beyond the critical period and persists, albeit in a diminished fashion, well into adulthood. Zeki refers to cases of dismemberment (like the loss of a finger) in which the mature brain of an adult person was able to reorganize itself such that the cortical area formerly assigned to the lost limb was relayed to take control of the remaining digits. He concludes that "this plasticity, perhaps on a more limited scale than during the critical period, is maintained well into adult life."[37] There is widespread agreement among neuroscientists on this point, which has been confirmed in numerous clinical trials and research studies over the past few decades.[38]

Human brain plasticity, and its persistence beyond the critical period into adult life, is of vital importance for the new field of neuroaesthetics and for bioaesthetics overall. In contrast to sociobiology and EP, which acknowledge that plasticity exists yet emphasize that our mental architecture is ultimately determined by genetic inheritance and epigenetic rules, neuroaesthetics shifts focus to the individual brain that takes up residence within that architecture, so to speak. "The microscopic variability of the brain at the finest ramifications of its neurons is enormous," the Nobel laureate Gerald Edelman points out, "making each brain unique."[39] This is not to deny that the basic development of the brain (such as its overall form, organization, and structure as well as its functional modularity and parallel operational mode) is genetically determined. Yet the neuronal specificity of each phenotype depends largely on its postnatal stimulation and interaction with the environment.[40]

Take learning, for example. Learning depends on memory, which in turn depends on the creation of cortical maps inside the brain. According to Hebb's law (formulated by the Canadian neurologist Donald Hebb in 1949), "cells that fire together, wire together." Hence, repeated exposure to the same stimuli over time will result in the creation of cortical maps that become "hardwired" in the sense of leaving permanent traces in the synaptic network of the brain. These traces constitute the physiological basis for memory. In neurological terms, to remember something is to reactivate a previously formed cortical map, which brings its representational content back to the level of consciousness. According to Nobel laureate Eric Kandel, this process of creating and reactivating cortical maps distinguishes long-term memory from short-term memory or working memory. The latter is based on experiences that occur only once and involve brief biochemical changes *within* individual cells (such as an increased release of neurotransmitters from the presynaptic to the postsynaptic cell). Long-term memory, by contrast, emerges only through repetition and involves changes in synaptic connectivity *between* cells. By paying attention to repeated stimuli of the same kind over extended periods of time, we allow the memory traces to become deeper and more richly embedded within the hippocampus (the memory center of the brain), where they can be reactivated later in life. Succinctly put: "Learning changes the strength of the connections among neurons."[41]

As far as we know, the overall cognitive power of the human brain depends less on its total amount of neurons than on the breadth and density of the synaptic network established between them. Hence a key question in contemporary neuroscience is what triggers the process of neuronal binding and enables the construction of large networks of interconnected cells across the brain. There exist two main explanatory models to date: the selectionist view (also called neuronal Darwinism) proposed by Gerald Edelman and Jean-Pierre Changeux, and the instructionist view favored by Joseph LeDoux and Eric Kandel. According to the former, the process of learning consists exclusively in weeding out unused neurons or failed synapses that crowd the system: "To learn is to eliminate," Changeux sums up the credo of the selectionist view.[42] More specifically, Edelman's theory is based on his earlier studies of the human immune system. The dominant theory until the 1970s had been that our immune system produces antibodies only in reaction to an intruding antigen. After the completion of its task, the newly created antibodies were believed to lie dormant until called on to combat a similar antigen in the future. But Edelman and other immunologists sought to demonstrate that immune recognition takes place through selection rather than instruction. Instead of building a specific antibody to meet a particular antigen, the body actually produces *in advance* a varied selection of different antibodies to combat any intruder it might possibly have to face later in life.

The basic idea behind Edelman's concept of neuronal Darwinism is to apply this immunological insight about the body's proactive generation of cell diversity to the interior workings of the human brain. The selectional theory is based on the fact that the amount of single neurons in a baby's brain peaks about three months after birth and then steadily declines throughout childhood and adolescence. According to Edelman, this means that humans are born with more neurons than they actually need, which requires the brain to keep eliminating neurons that have become dormant or no longer serve a purpose. In other words, "the brain, like the immune system, is a selection system that operates within an individual's lifetime."[43] Joseph LeDoux, by contrast, insists that "a purely selectionist view of the brain is not viable." He believes instead that learning also involves "an increase in synaptic complexity rather than merely a stabilization of the preexisting pattern—activity therefore is capable of instructing the formation of new synaptic connections." LeDoux's shift of emphasis from the elimination of neurons to the formation of new synaptic patterns explains why he dismisses Edelman's distinction between "developmental plasticity" and "learning" as more or less irrelevant: "Synaptic connections are adjusted by environmentally driven neural activity in specific neural systems. When these changes occur during early life, they are said to involve developmental plasticity; when they occur later, they are considered as learning. But the line between developmental plasticity and learning is a fine one and perhaps nonexistent."[44]

The difference between the selectionist and the instructionist model of how the brain works is obviously of tremendous importance for ongoing research in the neurosciences. Yet it is far less important in the context of neuroaesthetics, because both parties acknowledge and emphasize the enormous plasticity of the postnatal brain. The main point to keep in mind is that although our genetic heritage determines the overall framework for human brain development, the specific growth of synaptic connections within each individual brain depends mainly on postnatal life experiences and the particular kind of external stimuli it receives. Although this brain plasticity is particularly high during the so-called critical phase of infancy and early childhood, the neuronal structure of the brain nonetheless continues to evolve throughout adult life, albeit in a somewhat diminished and less versatile manner.

Consciousness

What does all of this mean for how scientists and philosophers today understand the nature of human consciousness? We saw in the first chapter that life can be defined as an autopoietic system able to interact with its environment. According to "embodied mind" theory (e.g., Varela, Maturana, Thompson), autopoietic systems of this kind are also cognitive systems, because they make distinctions and choices about how to adapt their internal structure to external perturbations. I also mentioned above that some biologists go a step further and actually assign consciousness to these systems as well: "Not just animals are conscious," Margulis and Sagan insist, "but every organic being, every autopoietic cell is conscious," meaning that "thinking and being are the same thing."[45] This view remains highly controversial among cognitive scientists, not least because "consciousness" is an inherently ambivalent term that carries along humanist connotations of "personal self" and "subjectivity" that are not necessarily implied in its biological meaning.[46] Most biologists readily concede that animals are conscious, yet also insist that only some animals—humans—are fully self-conscious. Dennett's view is paradigmatic for many in the field: "While a bat, and even the lowly lobster, has a biological self, it has no selfy self to speak of—no Center of Narrative Gravity, or at most a negligible one."[47] This categorical distinction, of course, also lies at the heart of current debates in bioethics and posthumanism, two fields that are highly critical of the reigning anthropocentrism not only in the life sciences, but across the humanities as well, as we shall see in the Coda.

Cognitive scientists have responded to this critique by proposing new conceptual models that serve to distinguish the highly developed level of human consciousness (evident in our sense of subjectivity and selfhood) from more basic, visceral–organic levels of consciousness common to all creatures.[48] While these conceptual distinctions have helped clarify the basic terms of the consciousness debate, they did not settle the longstanding philosophical dispute between the reductionist, third-person view of consciousness and the phenomenological, first-person view. The former is commonly referred to as "functionalism" or "eliminative materialism," because it explains the phenomenon of human consciousness solely in terms of brain activity. According to this view, mind equals brain. In the near future, Patricia and Paul Churchland prophesize, "we'll be able to say, as we did in the case of light or temperature, 'Aha, this is it. This pattern of activation in this context when the brain stem is doing such and such, that just is a sensation of red.'"[49] This position is shared by a large number of philosophers and cognitive scientists, including Daniel Dennett, Thomas Metzinger, and Gerhard Roth. The most concise formulation is provided by Francis Crick—one of the two Nobel laureates credited with discovering the DNA's double helix—in his "Astonishing Hypothesis": "The Astonishing Hypothesis is that 'You,' your joys and your sorrows, your memories and your ambitions, your sense of personal identity and free will, are in fact no more than the behavior of a vast assembly of nerve cells and their associated molecules. As Lewis Carroll's Alice might have phrased it: 'You are nothing but a pack of neurons.'"[50]

On the other end of the spectrum and opposed to the reductionist–functionalist view are those who insist on the irreducibility of consciousness to brain function. Their argument is that we cannot equate first-person experiences (i.e., the way it feels to be me) with third-person observations (i.e., the way in which my feelings appear to, or are interpreted by, another person). The "hard problem," according to David Chalmers, is *not* how to link a person's behavior or affective state to the electrochemical activation of specific areas in the brain. Rather, "the hard problem is the question of explaining how it is that all this is accompanied by subjective experience."[51] Supporters of this view, such as Stuart Hameroff, John Searle, and Roger Penrose, frequently point to qualia as the decisive part of a subjective experience that cannot be explained neurologically. They define qualia not in terms of particular attributes, but as the "intrinsic property of the experience itself," as something "private and ineffable."[52] To prove their point, analytical philosophers have invented a number of well-known thought experiments since the 1970s, such as Thomas Nagel's "What is it like to be a bat?" or John Searle's "Chinese-room experiment" or David Chalmers's "philosophical zombie."[53] These fictions are meant to demonstrate that one can well imagine some entity performing the very same tasks that we commonly associate with human consciousness *without*, however, that entity being conscious at all. They argue that consciousness, like life itself, can only be known from the inside. Our ability to study and comprehend autopoietic selfhood is inextricably intertwined with our own embodied being in the world. Evan Thompson puts it this way: "To make the link from matter to life and mind, from physics to biology and psychology, we need concepts such as organism and autopoiesis, but these concepts are available only to a bodily subject with firsthand experience of its own bodily life."[54]

Bioaesthetics, for sure, adopts this view as well. At the same time, I have no illusions that the philosophical debate about the nature of human consciousness will be resolved anytime soon, not least because the being of qualia remains just as elusive as that of memes: nobody knows what they really are. The only way out of this quandary was for everybody to agree on some neutral formulation that acknowledged the obvious links between mind and brain without privileging either side of the debate. Hence it has become customary over the last few decades to say that human consciousness "correlates" to neuronal activity in the brain. This hardly solves the problem, of course, for it remains unclear what, exactly, this means, given that each side draws different epistemological conclusions from the correlation—which, of course, was precisely why this formulation was deemed acceptable to both functionalists and phenomenologists in the first place. Does correlation actually amount to a real *causal* connection between brain activity and consciousness? And if so, is this *necessary* cause indeed *sufficient* to explain the phenomenon of consciousness? In my view, the answer to the first question is clearly "yes," much like the answer to the second question is clearly "no." Bioaesthetics rests on the premise that we cannot reduce mental processes to their physical carrier (i.e., mind to brain) without recognizing and studying the multiple levels of (conceptual and material) mediations in between the two. Mental properties supervene on physical matter, yet they are not reducible to it.

If one stresses scientific reductionism too much, then interdisciplinary conversations grind to a halt. This fact was made strikingly evident in a book-length dialogue between the neuroscientist Jean-Pierre Changeux and the phenomenologist Paul Ricoeur published in 1998. Much of their intriguing discussion centered on questions of terminology and linguistic meaning: how are we to formulate the relationship between the mental and the neurological, between the first-person versus third-person view of human nature? Here is how Ricoeur summarized his position: "I therefore propose that we say that the brain is a substrate of thought (in the broadest sense of the term) and that thought is the indication of an underlying neurological structure. Substrate and indication would thus constitute the two aspects of a dual relation, or correlation." Changeux, however, remained dissatisfied with this terminology:

> I do not see that your use of the term *substrate* illuminates the problem. In fact, I think it creates an ambiguity. Is it restricted to connectional anatomy? Why not then employ the descriptive phase *neural network*? Does it include activity or not? The discourse of the neurobiologist, which bears upon *three* distinct aspects—anatomical (neuronal connections), physiological (electrical activities and chemical signals), and behavioral and mental (action *on* the world and *internal* reflective processes)—seems to me much clearer.

In the end, Ricoeur and Changeux agreed to disagree. The overall problem, they concluded, "is really how to devise a uniform and consistent discourse" that enables scientists and philosophers to come to terms with their different views on the matter of consciousness.[55]

A main argument in this chapter, and throughout this book, is that such a uniform discourse does not exist—and that's a good thing, too. Because different disciplines employ different methods and tools that yield different objects of study. This applies even to disciplines that share the same methodology and claim to have the same referent in mind. The best example is the physical sciences. According to Werner Heisenberg, the single most important reason why his colleagues—including Albert Einstein, who until his death insisted that "God does not play at dice!"—were initially reluctant to accept the statistico-deterministic laws of quantum mechanics had to do specifically with the inadequacy of everyday language to express mathematical concepts: "The real problem behind these many controversies," Heisenberg concluded, "was the fact that no language existed in which one could speak consistently about the new situation." The real issue "did not concern the facts but rather the language."[56] The same, I believe, applies to the two cultures and the entire nature–nurture debate. One of the main reasons for the failed communication between scientists and humanists is that they do not share the same object of study even if they use the same term and seemingly refer to the same phenomenon, because each discipline shapes the object differently according to the particular conceptual framework that informs the study. From the scientific perspective, the conceptual weakness of traditional aesthetics resides in its failure to account for the evolved physiological constraints of human perception and aesthetic behavior. From the humanities perspective, the conceptual weakness of scientific approaches to art is their limited interest in—and knowledge of—the specificity of the art object *qua art*. While there is no objective middle ground in between these perspectives, bioaesthetics nonetheless strives toward methodological diversification and differentiation. The overall goal must be to avoid reducing the object of art to an object of science, and vice versa.

The only way to avoid this bifurcation, it seems to me, would be to use a universal language whose objectivity is devoid of semantics—that is, mathematics. In this sense only, I would agree with Alain Badiou's contested claim that mathematics is the ontology of being.[57] But mathematics' gain in conceptual clarity is paid for by the concomitant loss of embodied meaning, because mathematical concepts and symbols are (supposed to be) ideal entities. They have no immediate referent in the life-world of human beings. An n-dimensional space, for example, certainly makes sense *mathematically*, but it does not make any sense in terms of *human experience*. "What a paradox!" Edmund Husserl exclaimed in 1932 with regard to the empirical foundation of scientific knowledge since the Renaissance: "Nothing could cripple the peculiar force of the rapidly growing and, in their own accomplishments, unassailable exact sciences or the belief in their truth. And yet, as soon as one took into account that they are the accomplishments of the consciousness of knowing subjects, their self-evidence and clarity were transformed into incomprehensible absurdity."[58] Bioaesthetics acknowledges and builds on this paradox. It accepts the fact that there is no message without a medium, no observer without a perspective, and no organism without an environment. In short, "there is no description-independent way the world is, no way it is under no description."[59]

Neuronal Aesthetics: The Historical View

This premise, I think, informs the way in which artists and humanities' scholars conceive of the relation between neuroscience and aesthetics. Although the empirical approach continues to dominate neuroaesthetics overall, there also exists a minor history of defining the discipline primarily along aesthetic rather than scientific principles. As far back as 1978, the German philosopher and artist Bazon Brock gave a surprisingly forward-looking lecture on the "Neurophysiological Foundation of Each and Every Aesthetics"—an idea considered eccentric, even aberrant, by his audience at the time. Since then, Brock has repeatedly made the case for what he calls "neuronal aesthetics" *(Neuronale Ästhetik)*, an aesthetics based less on digital brain imagery and more on the critical assessment of scientific methods as such. "Neuronal aesthetics," Brock wrote in 1993, "seeks to understand the very conceptualization of neuronal processes as an aesthetic operation." As this quote demonstrates, neuronal aesthetics proposes to reverse the methodological premise of neuroaesthetics: rather than using scientific methods for the study of art and aesthetics, Brock suggests we should use aesthetic insights to better understand how contemporary science works: "Science today amounts to the production of signs and to dealing with these signs much like the arts have done for six hundred years," Brock argues. "Today, the motto is 'ut scientia poesis' (science, too, is a form of aesthetic production)."[60]

This does not mean that science is art. It merely suggests that science has much to gain from closer collaboration with the humanities, because science, too, must engage questions of semantics and meaning as part of its inquiry—an engagement that requires hermeneutic knowledge about the art of interpretation. Hans-Georg Gadamer, of course, had already made this argument back in 1960, and bioaesthetics reiterates it again today.[61] Brock's collaborator Olaf Breidbach, a biologist and historian of science, makes this point even more explicit. He strongly argues "against the idea that biological reductionism is helpful in addressing the question of aesthetics," because "aesthetics is not just the process of sensory imprinting, but also the process of *interpreting* such sensory imprints." Hence neuronal aesthetics, in Breidbach's view, must encompass "both the cultural and the physiological dimensions" of art and aesthetic experience. In explicit contrast to neo-Darwinian theory, Breidbach emphasized that neuroaesthetics implies, rather than contradicts, cultural relativism: "Neuronal aesthetics has shown the extent to which the categories we try to adopt are impacts of a particular culture." He concludes that "aesthetics, then, are not easily interpreted in neurobiological terms," because "culture is not genetically or neurologically determined."[62]

Mindful of this challenge, some recent approaches to neuroaesthetics have moved the individual art object back into the center of discussion. A pivotal question in the field of neuroarthistory is whether individual artworks not only reflect the increased plasticity of modern brains, but might have actively contributed to it. I shall address this question first with reference to Barbara Maria Stafford's work on baroque emblems and then with regard to Warren Neidich's theory of contemporary cinema and the "sculpted brain." Both approaches focus on brain plasticity, and both claim that the cognitive abilities of today's average brain far surpass those of its premodern ancestors prior to the Renaissance.

To avoid misunderstandings, it begs repeating that the basic anatomical structure of the human brain has not changed much over the last few thousand years. Yet the cognitive figures, schemata, and perceptual patterns that inform our experience of modern life certainly *have changed* considerably over the past few hundred years. As argued in the previous chapter, the cognitive plasticity of modern brains is primarily based on—and promulgated by—advanced modes of social learning and accelerated technogenesis.

An art historian with extensive training in neuroscience, Stafford, like most humanists, laments the fact that neuroaesthetics typically considers art a mirror rather than a window: it uses art mainly to exemplify rather than to augment today's scientific hypotheses about human evolution and cognition. Stafford's explicit goal, by contrast, is to establish a two-way street between art and neuroscience in order "to let art play a conceptual role in what neuroscience says about it."[63] Our phenomenological experience of and critical engagement with particular artworks, she claims, "becomes a dynamic type of nonlinguistic cognitive navigation."[64] Stafford's emphasis on the *nonlinguistic* cognitive power of images clearly distinguishes her account from the (persiflage of) postmodern aesthetic criticism derided by literary Darwinists and other evolutionary theorists. Stafford does not consider language the house of being, and she looks at artworks in order to *re*construct, rather than *de*construct, the modern mind. The recent advances in neuroscience, she claims, enable the humanities to move beyond the twentieth-century fetishization of language to rediscover the constructive cognitive power of images.

Stafford traces this inquiry back to eighteenth-century British associationism and European romanticism. The pre-romantics, she argues, were well aware that "nature and mind dynamically operate on one another" and constitute a "co-evolutionary process whereby the conditions occurring in culture and those happening in the brain are seamlessly linked together."[65] While this emphasis on coevolution—or what Stafford calls the "in-betweening" of concept and object, art and world[66]—echoes the basic tenets of sociobiology and evolutionary aesthetics, Stafford's theory moves beyond their broad anthropological framework by paying close attention to the historical particularity and media specificity of modern images. Referring to baroque emblems, for example, Stafford seeks to demonstrate that these artworks "permit us to see the synchronizing cerebral processes involved in vision."[67] They do so, Stafford claims, by activating the need for binding its disparate parts into a coherent, meaningful whole, much like the disparate synaptic connections in the brain must bind together to produce images in the mind: "Just as a coalescent self-awareness must continuously emerge from tangles of neurons, the activity of pictorial piecing involves an ongoing relational system."[68] Given this parallelism between sensory perception and neuronal binding, baroque emblems demonstrate how cognitive processes work precisely by engaging them as part of our aesthetic experience: "My overarching argument is that the . . . phenomenal attributes of a work of art are demonstrably also its primary means for manifesting evolving neural activity."[69] Hence artful images do far more than just mirror or activate pre-given perceptual patterns in the mind:

> They can also do something else. Additionally, they can conjure up the fact that something is creatively in the making. Representations of such incomplete situations are not merely "preconceptual." Rather, performative images or inlaying patterns let you perceive, as well as feel, how concept and world are constructed together. . . . I say thought is an image (since the different sensory modalities also generate "images") that incites us to re-perform (in the manner of mirror neurons) what we perceive.[70]

The study of modern art, in other words, facilitates our understanding of ongoing cognitive processes and epigenetic changes in the modern mind. I find this argument extremely compelling, and I believe Stafford's discussion of baroque emblems is exemplary for the kind of interdisciplinary perspective on art and human culture that informs bioaesthetics.

I am less convinced, however, by some of Stafford's more abstract and generalized conclusions about the alleged cognitive power of images sui generis, like her claim that "spatialized media . . . became, and continue to become, 'thoughtful.'"[71] I do not believe that bioaesthetics is best served by ascribing thought or desires to inanimate objects rather than locating it squarely within the human mind where it belongs. It is one thing to argue that some artworks both reflect and engage the ongoing process of neuronal binding inside the viewer's brain, yet quite another to imply that similar processes take place within art itself. There has been a plethora of similar claims made recently about *What Do Images Want?* (W. J. T. Mitchell), or about *How Images Think* (Ron Burnett), or *How Novels Think* (Nancy Armstrong), or *How Poems Think* (Reginald Gibbons), et cetera. To be sure, these are all substantive studies from well-established scholars in their fields, and I do not mean to question their legitimacy or overall acuity. Still, I believe bioaesthetics is better off avoiding this kind of rhetorical flourish regardless of whether it surfaces in neo-Darwinian theories of human culture or in humanities' studies of particular genres and individual artworks.

The danger of succumbing to this rhetoric is most apparent, I would argue, in neuroaesthetics' account of how postmodern "culture is sculpting the brain." The term "sculpted brain" was coined by the artist and cultural theorist Warren Neidich.[72] Like Stafford, Neidich explores the sociocultural ramifications of human brain plasticity as outlined by Nobel laureate Gerald Edelman and other neuroscientists (i.e., parallel processing, the copresence of input and output signals across the entire neocortex, and the dynamics of reentry). Yet Neidich shifts focus from early modern to postmodern art. Concentrating in particular on photography, film, and the new media, he claims that today's brains are radically different from their modern ancestors, because they are being shaped by the heightened "visual and cognitive ergonomics" of electronic media that adapt themselves "ever more adequately to the neuroanatomical and neurophysiological specificities of brain function."[73] Neidich's main argument is that most visual stimuli today are deliberately calibrated to increase the speed and efficiency with which our brains are able to absorb them. The most obvious example is the world of advertising, where skilled designers, psychologists, and market experts work together to craft powerful images specifically designed to please the audience's nervous systems and grab their attention. This positive feedback loop between visual stimuli and cortical structures creates an autocatalytic process that yields an ever-increasing overlap between external and internal patterns of cognition.

I agree with Neidich's central claim that "culture inscribes itself upon the brain." Given its ubiquity in contemporary culture, there can be little doubt that the medium of film continues to influence the construction of neuronal pathways inside our brains. Neidich, however, goes a step further. He stipulates the existence of what he calls a "'cinematic brain': a brain whose neural networks have been reconfigured by new time and space relations that define it."[74] Like Bernard Stiegler, Neidich believes that the cognitive effects of film extend far beyond the symbolic means of communication (e.g., language, facial gestures, bodily movement) used by narrative cinema to tell its stories. Instead, he emphasizes the importance of specific cinematic codes like montage or twenty-four frames per second, as well as film's auditory and proprioceptive effects on the viewer's nervous system. Given our constant exposure to film technology and images in today's mass culture, most viewers no longer consciously register the ubiquitous presence of these codes, which renders them all the more effective in rewiring the human brain. According to Neidich, this also helps to explain why the explicit goal of many contemporary artists is no longer to mirror or even intensify the ever-increasing speed of image production and reception, but to slow it down. The particular aesthetic achievement of new media installations such as Douglas Gordon's *24 Hour Psycho,* Neidich agues, is precisely to interrupt the visual ergonomics of postmodern culture by forcing spectators to take their time and re-familiarize themselves with the cinematic image.[75]

Media theorist Friedrich Kittler has made a similar argument with regard to the imperceptible yet ever increasing power of computer technology to shape the human mind. Kittler particularly lamented the technical illiteracy of most computer users today who use software applications or spend vast amounts of time online yet are unable to access, let alone comprehend, the complex programming codes that lie underneath. Such users, Kittler worried, increasingly become cognitively impaired and subject to easy manipulation by computer experts, because they fail to realize the degree to which meaning and cognition are being affected by digital culture and binary codes: "When meanings come down to sentences, sentences to words, and words to letters, there is no software at all. . . . All code operations . . . come down to absolutely local string manipulations and that is, I am afraid, to signifiers of voltage differences."[76]

The reductionist fervor evident in this passage and elsewhere in Kittler's oeuvre reminds me of today's biologism. While neo-Darwinians advocate for a Darwinian revolution in the humanities, Kittler advises us to rewrite the entire history of human culture in terms of mathematical symbols and binary codes. Both views blatantly ignore the many intermediate layers of materiality and meaning that operate between the (digital or genetic) source code and its measurable sociocultural effects. There is no straight line leading from one to the other. Instead, there is a complex structured field of intersecting and often contradictory forces with varying effects at different ontological levels that require detailed analyses every step of the way. The strength of Kittler's media theory rested precisely on his detailed historical analysis of new technologies and their epigenetic effects in modern society (see the prior chapter). Yet his penchant to abstract from these particulars in favor of general pronouncements about how electronic media control human behavior resulted in a kind of technological determinism strikingly similar to the genetic determinism I have criticized throughout this book. States Kittler, "Only when the cultural sciences begin to use contemporary [logarithms] to coordinate all the writing, reading and computing that history has seen will it have proved the truth of its renaming."[77] I sincerely hope this reductionism shall not come to define what Kittler calls "the cultural sciences" and I call bioaesthetics.

Neuroaesthetics: The Scientific View

Let us now examine more closely how scientists apply recent advances in neurophysiological research to the study of art and aesthetics. As Martin Skov, coeditor of a recent essay collection titled *Neuroaesthetics,* puts it, "neuroaesthetics is a new scientific discipline whose object it is to identify and understand the neural processes involved in human art behavior—those processes that underlie both the construction and experience of art."[78] I want to highlight the slight but important difference between this definition and the one provided in the introduction to the entire volume: "Neuroaesthetics is the study of the neural processes that underlie aesthetic behavior."[79] The first definition explicitly links neuroaesthetics to the study of art; the second does not. This difference testifies to a major conceptual problem in contemporary neuroaesthetics. For there is no compelling reason to link the physiological study of aesthetic behavior to works of art. Any other object, man-made or not, will do just as well. Indeed, ever since Duchamp's readymades entered the art scene some hundred years ago, aesthetic theory has had to acknowledge the fact that any given object can attain the status of art, depending on the social–institutional context in which it appears. The urinal Duchamp signed and submitted for exhibition in 1917 is commonly regarded as a work of art, whereas the exact same urinal located in a man's bathroom is not. Regardless of how we conceptualize the difference between these two objects, the crucial point is that this difference is *contextual and not material in nature,* because the physical nature of the object itself does not change regardless of whether it is considered a work of art. If we define neuroaesthetics strictly in terms of the neural processes that correlate to aesthetic experience, the formal qualities of individual art objects, as well as the institutional status of art *qua art,* becomes secondary and ultimately irrelevant.[80] But aesthetics is not synonymous with art. Many studies in neuroaesthetics continue to ignore this distinction and erroneously assume that their object of study is not just aesthetic experience or spectators' behavior, but also the phenomenon of art.

This problem is evident in an essay by the renowned physiologist Jean-Pierre Changeux titled "Art and Neuroscience" from 1994. In his essay, Changeux provided a detailed account of the neural mechanisms at work when an average person—let's call her the "implied spectator" analogous to the "implied reader" postulated by literary Darwinism—looks at a painting. To render his comments more concrete, Changeux chose to refer to a seventeenth-century painting (*The Lamentation on the Body of Christ* by Jacques de Bellange) housed in the Hermitage Museum in St. Petersburg. He writes:

> The analysis/reconstruction of a painting involves multiple areas of the brain in addition to those that specialize in the perception of colors. The neurons in these areas are mobilized together with those of the neighboring cortex, which are engaged in the task of analyzing the aforementioned attributes of form, distribution in space and movement. The analysis is completed by a synthesis. But whereas analysis may be a passive process, synthesis implies an "active" focalization of the viewer's attention. The painting as a mental object takes shape progressively.[81]

This is all true, of course. But it is no less true of the colorful flowers I see in my garden or the drawings of a five-year-old admired by her parents, because the neurological processes Changeux described are not specific to this particular painting nor to art in general. Changeux's analysis failed to reflect *in neurological terms* on the aesthetic specificity of this painting *as a work of art*. Doing so would require both a sociohistorical understanding of art and an account of the particular physiological effects that art—and only art—triggers in the brain.

Margaret Livingstone's discussion of Claude Monet's *Impression, Sunrise* from 1872 provides a good example of what I mean. Livingstone points out that the bright red sun in the center of Monet's painting is exactly the same luminance as the gray sky surrounding it. The reason it appears to viewers much brighter and more vibrant is due to the modular nature of our visual system and its two distinct neurological networks for processing light and color. From the earliest stages of the visual process that begins at the retina and leads all the way to the striate cortex and other areas in the brain, the qualities of light and color are analyzed separately. This disjuncture gives rise to the eerie effect of Monet's sun. For one part of our visual system (e.g., the rods in the retina and the luminance system in the brain) tell us the sun in this painting is no brighter than its immediate surrounding. In physical terms, this means that the amount of photons emitted by the painted sun is exactly the same as that of the painted sky around it. Yet the other part of our visual system (e.g., the retina's cones and the brain's color system) tells us that the sun is much brighter than the sky because of their different colors. This phenomenon is due to the fact that human color perception is more sensitive to light around 630 nanometers in wavelength (i.e., red) than a wavelength of 470 nanometers (i.e., blue). Hence red light appears far sunnier and more intense to us than gray or blue light of the same luminosity. Monet's painting skillfully exploits this discrepancy: both light and dark at the same time, his sun appears to pulsate right in front of our eyes.[82]

Many of today's popular media installations, like those created by the Icelandic artist Olafur Eliasson, achieve similar visual effects. Eliasson challenges viewers and critics alike to remain mindful of their embodied response to art. He does so by using digitally controlled signals (mostly images and sounds) that expose the involuntary mechanisms of human perception. Eliasson's explicit goal is to integrate the viewer's receptive process into the actual artwork itself: "I think that exposing the representational layer sort of clears the experience and makes it possible for us to see ourselves seeing—or knowing that we are seeing and seeing that we know."[83] Alva Noe has made a similar argument with regard to the huge sculptures of Richard Serra. Serra's work, Noe claims, "enables us to catch ourselves in the act of perceiving and can allow us to take hold of the fact that experience is not a passive interior state, but a mode of active engagement with the world."[84]

To be clear, I do not mean to suggest that the neurological effects explored in contemporary sculpture and media installations suffice to define the nature of art. Livingstone's study, however, does clarify how some artworks exploit the specific characteristics of our perceptual–cognitive system to evoke unique aesthetic experiences shared by all viewers alike. Obviously, this is not the norm, because most artworks have little or no control over how, when, or where we perceive them. This leads me to a second, related problem in neuroaesthetics. If the same object can function as art or not, depending on sociohistorical and institutional context, it follows that viewers' aesthetic behavior depends on this context as well. We all know from personal experience that the specific conditions under which art is produced, perceived, and evaluated influence our response to it. To claim otherwise is to assume that your (brain's) reaction to looking at a famous painting in the privacy of your own home would be exactly the same as standing in line for hours to catch a glimpse of it in the museum.

Or that it does not matter whether you contemplate the original *Mona Lisa* during a leisurely stroll in the Louvre or analyze a printed copy of it during your doctoral exam at the university. Why else, indeed, would forgery matter so much in the art world, if not for the fact that people's aesthetic behavior changes fundamentally once they realize they have been duped into admiring a masterful imitation as opposed to the original, even if only experts can barely tell the two apart? For neuroaesthetics to live up to its claims, it would have to measure the particular aesthetic–cognitive effects of such distinct experiences at the neuronal level in people's brains.

It is encouraging to see that some scholars in experimental neuroaesthetics have started doing precisely that. In his recent book *The Aesthetic Brain*, Anjan Chatterjee cites several recent studies designed to test the assumption that human bias, expectations, and previous knowledge co-determine our reaction to particular artworks. One such study by Kirk, Skov, and colleagues used fMRI scans to confirm that "people rated the same images as more attractive when they were labeled as museum pieces than when labeled as computer generated." Another study found that even a brief exposure to impressionist or cubist paintings along with some explanatory remarks about their art historical relevance sufficed to change viewers' neurological response when shown more images of the same kind.[85] Given the inherent limitations of brain imaging techniques, we certainly need to be careful not to interpret these results as conclusive evidence about how individual viewers will react to specific artworks. At the same time, however, I consider these studies important not only because they provide the kind of empirical data the humanities cannot produce on their own, but also because they clarify that neuroaesthetics has little to say about how culture shapes our understanding of art other than corroborate the fact that it does. Put differently: although neuroaesthetics can specify the way in which our sensations and emotions interact to produce aesthetic experience—specifications that will become increasingly relevant to bioaesthetics as the plasticity of our brains continues to increase due to cultural evolution and technogenesis—the new discipline cannot explain or elucidate the changing value or meaning of individual artworks embedded in particular sociohistorical contexts. Chatterjee makes this point explicit: "Scientific aesthetics can scrutinize general effects of knowledge on aesthetic encounters, but not the specific knowledge and layered meanings woven into individual works of art."[86]

This insight has moved to the center of the debate in neuroaesthetics over the last few years. The crucial point is that neuroaesthetics can be applied to any object and not just art objects, and that the knowledge it produces is of limited value for the kind of research conducted in the humanities. Skov and Vartanian clearly struggle with this dilemma at the heart of neuroaesthetics: "Ultimately, we will need models that can account for our aesthetic experiences with all objects rather than just artworks, although the field's current focus on the latter is a justifiable starting point."[87] Yet it remains unclear what that justification could possibly be, given that much of neuroaesthetics continues to disregard the sociohistorical dimension of art production and reception. Instead, their empirical approach harks back to the history of nineteenth-century physiological–experimental studies of human perception, as the editors' explicit reference to Gustav Theodor Fechner's *Vorschule der Ästhetik* from 1876 confirms.[88]

On the one hand, this is good news, because it signals that neuroaesthetics starts to overcome the historical amnesia and futurist-oriented approach that characterizes modern science in general and the neurosciences in particular. On the other hand, however, there is little recognition in the field that Fechner's study merely sought to provide aesthetic discourse with relevant empirical knowledge about the biological *constraints* of the human sensorium as opposed to its alleged *nature* or *essence*. Fechner's critique was directed against the intellectual hegemony of idealist philosophy in nineteenth-century German aesthetics, and he tried to reestablish an intellectual balance between empirical science and speculative interpretations of art. Fechner's title (*Vorschule* literally means *preschool*) clearly indicates that his "aesthetics from below" was not meant to replace, but rather to augment the reigning "aesthetics from above."[89] It was self-understood at the time that no discipline—no matter how comprehensive or objective it proclaims to be—could possibly bridge the epistemological gap that separates our scientific knowledge of biological processes from aesthetic experience in general and our affective–cognitive response to art in particular.

Fechner's relatively modest goal to complement rather than replace traditional aesthetics has given way to the far greater ambitions of scientists today. This was made evident in a highly influential article from 1999 titled "The Science of Art: A Neurological Theory of Aesthetic Experience." Cowritten by the neuroscientist V. S. Ramachandran and the philosopher William Hirstein, the essay started a series of interdisciplinary debates on the relation between art and the brain published in the peer-reviewed

Journal of Consciousness Studies between 1999 and 2004. These debates not only clarified the different methodologies used by scholars from different fields. They also revealed the deep distrust and mutual suspicion that divides scholars in favor of empirical, scientific studies of art and aesthetics from those who insist on the hermeneutic, art historical study of individual objects, genres, and cultures. This rift confirms the basic lesson learned in the prior chapter, about literary Darwinism: the effort to enforce consilience (to use E. O. Wilson's term) between academic disciplines inevitably results in epistemological gridlock and moralizing rhetoric.

As is typical for neuroaesthetics in general, Ramachandran and Hirstein's study disregards the discursive–institutional dimension of art, an omission the authors explicitly address early on in their essay: "Indeed," they admit, "it may well be that much of art really has to do with aggressive marketing and hype, and this inevitably introduces an element of arbitrariness that complicates the picture enormously."[90] But their dismissive tone makes it all too clear that the authors fail to understand the crucial importance of this "complication" for their own research. For outside and beyond what they call the "aggressive marketing and hype" of art—that is, outside the institutional space and discursive network of art markets and art criticism—the phenomenon of "art" simply ceases to exist. By contrast, Ramachandran and Hirstein argue that "there are universal rules or principles" for visual art, and that these rules can be revealed with the help of the neurosciences. Dennis Dutton, we recall from the previous chapter, made a similar claim based on evolutionary theory, except that he came up with a list of eight defining characteristics of art, whereas Ramachandran and Hirstein come up with their own list of twelve essentials.

One of these rules is called the "peak shift effect," which is another term for what Wilson calls the "supernormal stimulus."[91] Researchers have found that if a rat gets rewarded for selecting a particular shape (say, a rectangle) over another shape (say, a square), then the rat will "respond even more vigorously to a rectangle that is longer and skinnier than the prototype." In other words, the rat will recognize the distinctive characteristics of a rectangle and react positively to a shape that further highlights these characteristics. Ramachandran and Hirstein's basic idea is to apply this neurological principle to the perception of art. "All art is caricature," they claim, because art, like caricature, aims at highlighting some features at the expense of others. "Given constraints on allocation of attentional resources, art is most appealing if it produces heightened activity in a single dimension (e.g., through the peak shift principle or through grouping) rather than redundant activation of multiple modules." The reason we react more strongly to such single-dimensional art is that it "titillate[s] the visual areas of the brain" and more quickly establishes a connection to the limbic system associated with the pleasure of having "recognized" a pattern or object. So, the easier art makes it for us to recognize or discover something worthy of our attention, the more pleasurable and enjoyable the experience of doing so becomes. The explicit premise is that art serves some practical purpose and has a "survival value" for the species. This premise, too, echoes the evolutionary–utilitarian argument already encountered in the prior chapter. It essentially reduces the history of art to a succession of neurological training camps of sorts, the evolutionary function of which is to increase efficient pattern- and object-recognition in order to better detect enemies, pursue prey, and adapt to the environment. The "overall function of art," in other words, is "the acquisition of knowledge."[92]

It is hardly surprising to find that critics' responses to Ramachandran and Hirstein's essay followed the same disciplinary divide we encountered in our previous discussion of literary Darwinism. While most natural and social scientists offered various points of critique but explicitly praised the innovative character of the essay, the humanists' reaction was far more critical.[93] Two responses are particularly noteworthy. The first is the exceedingly short and dismissive paragraph by the famous art historian E. H. Gombrich, who ends his comments with the following suggestion: "Even a fleeting visit to one of the great museums might serve to convince the authors that few of the exhibits conform to the laws of art they postulate"—the implication being that Ramachandran and Hirstein's definition of art misses its target entirely and does not warrant any further discussion.[94] This curt rejection of Ramachandran and Hirstein's approach may seem surprising, given that they, along with Changeux and others, have explicitly praised Gombrich's work as inspirational for their own studies on neuroaesthetics. Yet Gombrich's seminal book on *Art and Illusion* (published in 1960) studied the psychology of perception with regard to the *historical* modes of artistic representation as opposed to their ahistorical, *essentialist* core. His primary focus was the single work of art, not the mental architecture of the brain. Although Gombrich certainly made use of contemporary experimental research about vision, he did not provide a physio-neurological account of the inner workings of the brain. After all, behaviorism was still a major force in psychology at the time, and it was only later (in the 1970s) that the newly formed discipline of cognitive science began to concentrate on the interior workings of the mind. Most important, Gombrich did not equate aesthetic experience with either pleasure or evolutionary fitness, as Ramachandran and Hirstein ended up doing.[95] The obvious problem with this equation is that it reduces the *sensus communis aestheticus* to the rationalist notion of common sense (see chapter 1). This approach falls far behind Kant's efforts to develop a genuine aesthetics distinct from both pleasure and cognition.

The second response I want to mention is the brazen critique by Donnya Wheelwell, who ranked Ramachandran and Hirstein's study "among the most clear-cut examples of late twentieth century sexist scientific and reductive megalomania."[96] However one may judge the academic merits of this harsh review, one should note that the name "Wheelwell" was actually a pseudonym meant to protect the real identity of its author, as the journal later pointed out—an admission that led Ramachandran to deride Wheelwell's "intellectual cowardice" and to protest against the editorial policies of the journal. Taken together, Gombrich's silence and Wheelwell's excessive rhetoric speak volumes about the larger issues at stake—namely, the dismal academic climate and the mutual suspicion of intellectual dishonesty that continues to separate the two cultures. If nothing else, the Ramachandran/Hirstein debate served to expose the lack of a common language able to bridge the gap between the two cultures and facilitate a productive dialogue between them.

"The Pre-existing Idea within Us": Zeki's Platonism

Apart from Anjan Chatterjee, the scholar most poised to accomplish this feat is Semir Zeki. There are several reasons for Zeki's preeminence in the field. Most important, Zeki's version of *neuroesthetics* is equally informed by neuroscience and by art history, and his detailed interpretations of particular artworks range from the Renaissance to twentieth-century abstract paintings. Second, Zeki wisely remained agnostic about what constitutes the "nature" of art. Unlike many of his colleagues, Zeki does not elaborate art in terms of aesthetic behavior (as do Dissanayake, Coe, Skov, and Vartanian), nor does he seek to define its essence along eight or twelve or any other number of allegedly indispensible characteristics (as do Dutton, Ramachandran, and Hirstein). Zeki simply accepts the existence of art as a given, which allows him to sidestep the heated debate about cultural relativism versus universalism that continues to dominate evolutionary aesthetics.[97] Rather than trying to define *art*, Zeki defines the *function* of art. The function of art, like that of the brain, is to produce knowledge. Art, Zeki argues, seeks "to represent the constant, lasting, essential, and enduring features of objects, surfaces, faces, and so on, and thus allow us to acquire knowledge."[98]

This focus on the cognitive function of art means that Zeki's neuroesthetics resonates well with other strands in evolutionary aesthetics, including literary Darwinism and cognitive cultural studies. Human knowledge, Zeki insists again and again, is a product of the brain, whose neurological structures result from the (phylogenetic and ontogenetic) history of external stimuli processed over time. Knowledge results "from the many images that [the brain] has seen, and from its ability to select from those images only that which is necessary for it to extract the essential qualities of objects." The more stimulation the brain receives, the more discerning it becomes. It follows that the essential contribution of art to knowledge, in Zeki's view, is its ambiguity—ambiguity understood not in the sense of total uncertainty, but rather as "the certainty of many different, and essential, conditions" that constitute the human environment.[99]

The distinctiveness of Zeki's theory is most evident in comparison to that of Ramachandran and Hirstein. Both parties emphasize that art produces knowledge, that art is pleasurable, and that the two are related: art is pleasurable because it produces knowledge. But whereas Ramachandran and Hirstein provide a broad anthropological survey of cultural artifacts to support their central thesis that the function of art, like that of caricature, is to *reduce* the complexity of the world, Zeki, by contrast, focuses on Western canonical paintings since 1600 to support the opposite claim: art is pleasurable and produces knowledge because it *foregrounds* and thus forces the brain to *think through* the complexity of the world. The aesthetic power of Vermeer's paintings, Zeki argues, is "its capacity to evoke many situations, not one, all with equal validity and hence to cover a 'whole species of situations.' It has the capacity to stir a great deal in the brain's stored memory of past events."[100]

From a bioaesthetic perspective, Zeki's approach is far more interesting than that of Ramachandran and Hirstein, because it allows for in-depth analyses of individual artworks. At the same time, we should note that the two approaches are not mutually exclusive. There is no reason to deny that some artworks are pleasurable because of their simplicity, while others are pleasurable because of their complexity. Yet a good many of them, I would argue, are not pleasurable at all. Or they might be pleasurable for some people, but not others, or pleasurable today, but not tomorrow. Some works are not beautiful, but are meant to be ugly, hideous, or repulsive. My point is that neither cognition nor pleasure provides the necessary, let alone sufficient, criteria for defining aesthetic experience, not to mention (the cognitive function of) art. Instead, this view reduces aesthetic judgments to either cognitive judgments or judgments of taste, meaning that neuroaesthetics circles back to the precritical philosophical tradition Kant's *Third Critique* had already left behind.

Finally, I would argue that Zeki's exemplary status in the field results from the balanced view he advocates with regard to brain modularity versus brain plasticity. On the one hand, Zeki's empirical research clearly confirms the innate modular structure of the brain, which features distinct areas primarily responsible for specific tasks such as vision, hearing, memory, et cetera. On the other hand, Zeki readily acknowledges that the process of neuronal binding (both within and between different cortical areas) has not yet been sufficiently understood, and that there are simply too many parallel processes going on at the same time to conceptualize them solely in modular terms. As he states in *Vision of the Brain*:

> Thus, while there is no doubt that the cerebral cortex is modular in its organization in the sense that, even for a given modality, it assigns geographically separate cortical areas and sub-regions of areas to different specializations, we still have no agreement as to whether there is a unit of organization which is repetitive, and the same everywhere in the cortex, and which we can think of as the cortical module.[101]

Although Zeki concedes that the neurosciences have yet to develop a precise concept of "the cortical module," he is nonetheless convinced that future research will provide conclusive evidence of its factual existence. This belief, too, is paradigmatic of evolutionary aesthetics at large. In fact, we have already encountered a similar faith in scientific progress at various points throughout this book—like Schrödinger's belief that scientists will soon discover new laws in physics to explain life, or Wilson's assumption that molecular genetics will soon be able to define human nature in scientific terms, or Dissanayake's and Carroll's claim that evolutionary theory will soon revolutionize and unite all humanities research under the paradigm of consilience, et cetera. The pervasiveness of this faith throughout the scientific community, I argue, does more than expose the hubris of modern science. It also reminds us that scientific research frequently operates with highly speculative concepts (like genes or memes, culturgens or cortical modules) whose material referent is presumed rather than proven to exist. There is nothing wrong with such speculations, of course—provided they are recognized and remembered for what they are. Otherwise, there is the danger that, sooner or later, these concepts will evolve into entities as dubious and elusive as Hegel's spirit or Plato's ideas.

The latent idealism of contemporary science, indeed, comes to the fore in Zeki's version of neuroesthetics. To support this claim, I want to focus on Zeki's discussion of abstract art, in particular chapter 13 of *Inner Vision* titled "Mondrian, Ben Nicholson, Malevich, and the Neurophysiology of Squares and Rectangles." This chapter explores the formal symmetry between the vertical and rectangular lines featured in cubist and suprematist paintings, on the one hand, and the rectangular shape of the receptive field of neurons in the visual brain, on the other. Depicting both shapes next to each other on the same page, Zeki comments on them as follows:

> I think that the similarity between the two, the receptive field structure and characteristics of a cell on the one hand and the creations of artists such as Malevich on the other, is really quite striking. This relationship is made all the more compelling when one reflects that the painting is the creation of a brain that contains cells with this kind of receptive field described above.[102]

While the formal similarity is undeniable, we should remain mindful about what, exactly, this comparison entails. In his book, Zeki provides a photo reproduction of Malevich's famous painting *Red Square,* and right above it, he depicts a graphic representation of the firing rate of neurons in the visual brain. This firing rate, Zeki explains, is determined by the functional specialization of individual cells, which react only in response to certain kinds of stimuli (e.g., specific movement, color, form) and only within the particular spatial parameters that constitute their receptive field. Exposed to the wrong stimuli and/or exposed to stimuli outside their receptive field, these neurons will not react. And due to their center–surround sensibility, they will also not react, or react less vigorously, in the absence of strong stimuli differentials at the edge of their receptive field. The crucial point is that these receptive fields, albeit varying in size, "are usually square or rectangular in shape" *regardless* of the specific input selectivity (e.g. color, shape, movement) of the cells involved.[103]

Zeki's comparison highlights the similarity between these two squares—Malevich's *Red Square* and the shape of the receptive field of neurons in the visual brain. In my view, however, the two are like apples and oranges. For unlike Malevich's painting, the graph does not actually *represent* a square; instead, it signifies the firing rate of a particular neuron exposed to appropriate stimuli within its receptive field. That field, to be sure, forms a square, a point nicely illustrated by using the graph's vertical lines to represent both the cell's firing rate and the rectangular shape of its receptive field. Yet the empirical data—that is, the amplitude and frequency of neuronal activity measured with the help of electronic apparatuses—remains the same regardless of how we chose to represent it. The reason for this substitutability of symbolic codes is that modern science considers symbols extraneous to the data they represent. A row of numbers, for example, would look nothing like a square yet still signify the neuron's receptive field properties just the same as the graph. Obviously, the same cannot be said about Malevich's representation of the *Red Square,* because it has no referent outside itself. Being a work of art, its signifier and its signified are one and the same—namely, this particular red square here and now. Malevich's red square is a haecceity, to use Husserl's term: a unique singularity in space and time that ultimately signifies nothing beyond its own being. This distinguishes it from the generic rectangular space outlined by the receptive field of neurons in the visual brain.

To clarify this difference, let us think back to our earlier discussion of digital brain imagery. Functional MRI technology, we noted, does not actually *take* a picture of the brain but *makes* a picture of the brain by translating numerical data about cranial blood flow into a computer-generated image that features a bunch of red dots. Much as these brain scans do not actually represent a brain, even though they are made to look like one, the receptive field of neurons in the visual brain does not actually *represent* a square, even though it happens to coincide with that geometrical form. Put differently, even if we had a magical microscope that allowed us to look directly into the visual brain, we would not see a single square of the kind Zeki talks about, because they do not exist physically in the brain. The square-shaped receptive field of vision is an abstract entity whose existence is inferred from empirical data, not material forms.

Zeki, in my view, all too readily collapses these two different kinds of squares into one: "If one were to view the Malevich painting from a distance that is sufficiently large," he points out, "the entire square in the Malevich painting could fall onto the receptive field of a single cell like the one illustrated."[104] That is true. Yet apart from this spatial congruence, there is no other significant connection between these two squares. The receptive field properties of neurons in V4 remain the same regardless of whether these cells are exposed to Malevich's *Red Square* or to any other similar object chosen at random. So what does this similarity signify in either neurological or aesthetic terms? Zeki struggles to provide a compelling answer to this question: "One would be foolish to equate the two," he readily admits, because one "cannot draw an exact causal relationship between the two."[105] To his credit, Zeki—unlike some of his colleagues—never claims that the *Red Square* actually represents or signifies a receptive field, nor does he suggest that Malevich or Kandinsky ever intended to paint such a field.[106] Zeki is nonetheless determined to hold fast to the similarity he has discovered: "I find it difficult to believe that the relationship between the physiology of the visual cortex and the creations of artists is entirely fortuitous."[107] Hence he concludes "that the Malevich work would not produce any aesthetic effects but for the presence of these cells, which is not the same thing as saying that these alone produce the aesthetic effects."[108]

Zeki's ambivalence on this point is telling, because the only philosophical framework in which his comparison makes any sense at all is Platonic idealism. One would have to assume that both forms—Malevich's painted square and the rectangular-shaped receptive field of neurons in the visual brain—emerge from, or signify toward, some universal form of "rectangularity" as such. This, indeed, is how I understand Zeki's main conclusion at the end of this chapter: "The fact is that the new forms [of cubism], consisting largely of lines, squares and rectangles, are admirably suited to stimulate cells in the visual cortex, and the properties of these cells are, to an extent, the pre-existing 'idea' within us."[109] If my reading is correct, then how does Zeki hope to reconcile his seemingly idealist view of art with the empirical nature of neuroscientific research—particularly in light of the fact that elsewhere in his book, he explicitly criticizes Plato's anti-empiricism as incompatible with neuroaesthetics?[110] One possible option to bridge the tension between the two would be phenomenology, clearly one of the most influential attempts in twentieth-century philosophy to transcend the binary of idealism and materialism. Yet Zeki makes no mention of either Husserl or Merleau-Ponty, nor does he refer to current cognitive theories about enactment or the embodied mind (e.g., Varela, Thompson, Lakoff, Johnson).

Instead, Zeki does what the best scientists, according to Louis Althusser, have always done since the seventeenth century: he develops an SPS, a "spontaneous philosophy of the scientists."[111] I hasten to clarify that this claim is not meant to belittle Zeki's admirable knowledge in the history of philosophy and art. On the contrary, I agree with Althusser that only the best brains in the sciences are able to divine an SPS at all. Doing so requires not only broad interdisciplinary knowledge across academic fields, but also the intellectual aptitude and academic courage to interpret scientific data in philosophical terms. In Zeki's case, his SPS tries to historicize Plato via Hegel: he identifies both Hegel's "concept" and Plato's "idea" with "the brain's stored record" of past experience:

> Whether art succeeds in presenting the real truth, the essentials, or whether it is the only means of getting to that truth in the face of constantly changing and ephemeral sense data, the opposing views are at least united in suggesting that there is (Hegel) or that there should be (Plato and Schopenhauer) a strong relationship between painting and the search for essentials. And my equation, of both the Hegelian Concept and the Platonic Ideal with the brain's stored record, means that the difference between the two, from a neurological point of view, is insignificant.[112]

As I see it, Zeki's neuroaesthetics is a brilliant attempt to enrich the empirical world of scientific materialism with the humanist spirit of philosophical idealism. This union certainly makes sense from Zeki's scientific perspective, but it makes far less sense from a bioaesthetic perspective. For there is obviously a significant aesthetic difference between Plato's assumption that truth is absolute and can never manifest itself as such in art, or Hegel's view that art has (or will) become obsolete once spirit becomes absolute, or Adorno's belief that truth negates itself in and through the modern work of art, or Deleuze's claim that art and science merely create the semblance of truth via the production of concepts, percepts, and affects. Bluntly put, from Plato and Hegel to Adorno and Deleuze, the constructivist notion of truth becomes stronger while its essentialist notion grows weaker.

It is difficult to locate Zeki's theory on this philosophical spectrum, however. Although he provides scientific evidence for the constructivist nature of human cognition, he also remains committed to the idealist belief that science and art provide knowledge of essential forms and timeless being, which affords us a glimpse of truth itself. But if knowledge emerges from historical interactions between human brains and their environments, would it not be more apt to say that art and science serve to *produce* knowledge? This formulation, I think, bridges the gap between Zeki's claim that "art is a search for constancies" and the opposite claim articulated by John Dewey back in the 1930s: "The conception that objects have fixed and unalterable values is precisely the prejudice from which art emancipates us."[113] From a bioaesthetic perspective, both propositions make sense. It is precisely because art challenges traditional cognitive structures and preconceived notions of truth that it creates different cognitive patterns within the brain to help us detect new consistencies in our environment.

Rather than harking back to Plato and Hegel, the "pre-existing idea within us" might better be understood in Deleuzian terms as a virtual space of becoming that is actualized via a process of intensive differentiation.[114] A key objective of Deleuze's philosophy, after all, was nothing less than "to reverse Platonism" with the help of science.[115] "The idea," writes Deleuze, "that truth is not pre-existent, something to be discovered, but instead, must be created in every field, is easily seen in the sciences."[116] Neuroaesthetics, I think, aptly demonstrates that biology and the sciences in general partake of this constructivist notion of truth—the only kind there is. Human knowledge and our sense of truth vary historically because they depend on the methods and concepts we use to enact the world we live in. Hence Deleuze kept emphasizing the productive dimension of art and aesthetics. Aesthetics is indeed a "*science* of the sensible," he argued, yet it must not be misunderstood as a mere *representation* of what already exists. Instead, aesthetics helps to *produce* human sensibility and cognition in the first place.[117] Aesthetics, for Deleuze, becomes scientific "only when we apprehend directly in the sensible that which can only be sensed, the very being of the sensible: difference."[118] In that sense, bioaesthetics is Deleuzian through and through.

CODA

Living is sense making

FRANCISCO VARELA

Deleuze and Affect

"The enemy is the organism."[1] The organism prevents the actualization of the body's virtual power of becoming: "What a body can do is the nature and the limits of its power to be affected," Deleuze famously stated in his book on Spinoza, and he added, "*We do not even know of what affections we are capable, nor the extent of our power.*"[2] In order to unleash this power, the human body must shed its organs and its organization. It must transcend both the evolutionary history of its species (phylogenesis) and its acquired habits during life (ontogenesis). The mind that pertains to this body is unmoored from consciousness and transcendental subjectivity—the philosophical tradition that runs from Descartes to Husserl—and from the phenomenological experience of being human studied by John Dewey and Merleau-Ponty, Varela and Lakoff. Instead, the body without organs becomes subject to the affects that run through it. Affects, according to Deleuze, are synonymous with the vital, virtual forces of pure becoming that "inhere" or "subsist" within the actualized reality of organic life. Their creative power exceeds human being and its lifeworld. "*Affects,*" Deleuze and Guattari emphasize again and again, "*are precisely these nonhuman becomings of man.*"[3]

The attempt to overcome the biological constraints of human corporeality separates Deleuze and Guattari's aesthetics from the notion of bioaesthetics developed in this book. Bioaesthetics, I want to emphasize again, is informed by the experience of being human and the history of human beings, whereas Deleuze's aesthetics of affect is informed by our potential to become other than human. Between the two, the latter is dominant today. Even a cursory look at current scholarship confirms that Deleuze's vitalism is the single most influential source of the "affective turn" that has taken hold of the posthumanities in recent years.[4] From a bioaesthetic perspective, Deleuze's influence has been a mixed blessing. On the one hand, his vitalist philosophy has led many to conclude that affect is autonomous and completely severed from cognition.[5] That is a mistake. On both the phenomenological and the physiological/neurological levels, affect and cognition intertwine, and it is impossible to say where sensation ends and cognition begins. Affects, for sure, are not the same as conscious emotions. Yet recent studies in affective neuroscience have confirmed that the functional differentiation between lower and higher forms of human cognition includes the reciprocal exchange of information back and forth between them. In the words of Jaak Panksepp, the author of *Affective Neuroscience*, "the interchange between cognitive and emotional processes is one of reciprocal control."[6] Much like our sense organs always already "interpret" the external stimuli they receive before processing data to the brain, the limbic systems both *sends and receives* internal signals *to and from* the neocortex. And much like there is no single cortical module for feelings or emotions, there is no affective state entirely severed from cognition: "Rather, affect and cognition are inextricably engaged in all brain and mind processes, creating meaning out of information," Steven Rose points out.[7]

The constitutive intertwinement of affect and cognition is not always appreciated, and sometimes ignored, in affect studies.[8] Instead, there has been a tendency to shift focus from theory to rhetoric or to sidestep methodological questions altogether. The editors of *The Affect Theory Reader*, for example, argue

> that, in theory, the "what" of affect often gives way to matters of "how" in the rhythm or angle of approach: thus, why a great many theories of affect do not sweat the construction of any elaborate step-by-step methodology much at all, but rather come to fret the presentation or the style of presentation, the style of being present, more than anything else.[9]

This switch from methodology to style, I fear, will lead affect studies right back to the speculative poetics that has dominated Western theory for most of the twentieth century. It looks like a dead end to me. How many more master stylists like Heidegger or Derrida do the humanities need? And how much more aestheticized can theory get before it reverts back to Adorno's aesthetic theory, a text written some fifty years ago?[10] I think the affective turn got it exactly backward: bioaesthetics requires *more* methodological reflection, not less. Otherwise, it risks "the affective fallacy" already denounced by the New Critics: "The Affective Fallacy," Wimsatt and Beardsley argued, "is a confusion between the poem and its *results* (what it *is* and what it *does*)," the outcome being "that the poem itself, as an object of specifically critical judgment, tends to disappear."[11] Many affect studies published over the last decade are subject to this charge: they lose sight of their object of study.

Ostensibly directed against the idealized notion of human subjectivity, the affective turn has succeeded to do something else entirely, namely, to undermine a coherent sense of emotion and cognition as coevolving traits of being human. And it has done little to help clarify the conceptual confusion about the nature of emotions across the disciplines.[12] There are, luckily, numerous ways we can think about affect beyond Deleuze's vitalism.[13] And there is a rich academic history of distinguishing affect from human feelings and emotions.[14] Unless we conceptualize affect in the context of current evolutionary and neurological research about the human body, we run the risk of mystifying its effects in the name of some unknowable force that literally moves the world entirely beyond our control. Such a vitalist notion of affect becomes indiscernible and inexpressible—a metaphysical construct devoid of the transformative social and political power often attributed to it.

On the other hand, Deleuze's influence in the (post)humanities has also prompted numerous studies that recognize and integrate the biological dimension of affect into their framework. Although Brian Massumi, for example, emphasizes that affects themselves are "not ownable or recognizable and . . . thus resistant to critique," he nonetheless insists they are "analyzable in effect, as effect." Although affects themselves cannot be perceived and made conscious, this very fact (i.e., that affects cannot be perceived) *can* be made conscious. We know that affects escape us because we can perceive the effects of our inability to perceive them. What is needed, therefore, is "a vocabulary . . . for that which is imperceptible but whose escape from perception cannot but be perceived, as long as one is alive."[15] I agree. Yet, unlike Massumi, I see no reason why this approach should be judged incompatible with traditional hermeneutics or the phenomenology of aesthetic experience. One may grant the unknowable nature of "pure affect" in the Deleuzian sense and then move on to examine its physiological and material *effects* in relation to politics and human culture.

This, precisely, is what the best studies of affect—works by John Protevi, Manuel DeLanda, William Connolly, and Lauren Berlant, among others—are doing today. Their focus rests neither on the virtual forces of becoming nor on the actualized stratifications of being. Instead, they focus on the dynamic thresholds in between—that is, the particular points at which affect seizes human bodies and causes them to (re)act differently. What matters to them is the sociocultural transmission of affect—the way in which affect travels from one person to another, from the artwork to the spectator, or from the social environment to the human individual and back again. These thresholds and transmission processes, however, are *both constructed and real.* They are not only material, but also conceptual in nature, and there is no sense in arguing about which aspect is primary or more important. Rather than juxtapose cognition and affect, as many other Deleuzians have done, Protevi provides a wonderfully nuanced and historically contextualized account of how the two interact and shape the lifeworld of human beings.[16] And rather than denounce human subjectivity tout court, Berlant identifies the need to construct new forms of subjectivity as a key topic in contemporary literature.[17] Rosi Braidotti, too, acknowledges the growing political demand for what she calls *"posthuman* subjectivity." Although Braidotti defines this subjectivity in Deleuzian terms as a relational, non-centered assemblage informed by "the vital, self-organizing and yet non-naturalistic structure of living matter itself," she also emphasizes the need "for self-reflexivity on the part of the subjects who occupy the former humanist center."[18] The paradox, in other words, is that we need some kind of subjectivity to think beyond it and still act responsibly. Posthumanism must *work through* humanism rather than ignore it altogether. This ethical commitment clearly distinguishes bioaesthetics and critical posthumanism from both affect theory and the flat ontology espoused by so-called speculative realism and object-oriented ontology.

Deleuze's own philosophy easily accommodates both the poetic and the methodical turn in current affect theory, albeit unevenly. On the one hand, there are numerous passages throughout his oeuvre where he attributes a cognitive dimension to affect. "Thinking is thought through . . . sensations," he claims,[19] and "there is as much thought in the body as there is shock and violence in the brain."[20] Deleuze emphasized again and again that actuality and virtuality, though distinct, are intertwined ontologies at the intensive level. On the other hand, however, Deleuze's concepts all too readily withdraw from the actual into the virtual. Though he claims there is thought in affect and sense in sensation, he also claims that sense remains autonomous and irreducible to human sensations. Similar to affect, sense, for Deleuze, is "pre-individual, non-personal, and a-conceptual."[21] Most important, Deleuze insists that "the logic of sense" has nothing in common with common sense or the *sensus communis*. It is an entirely new sense apart from the evolutionary history of the human species. That explains why "the people are missing," as Deleuze puts it. There is sense in the world, but not for us, only for "the people to come."[22]

A Posthuman Aesthetics?

Bioaesthetics, too, promotes a new sense of human being and subjectivity. Yet it believes this sense must be rooted in, and emerge from, our shared nature and history of being human. Deleuze's frequent derision of "common sense" and "good sense" shared within and among human communities is the most disconcerting aspect of his philosophy, in my view. It gives rise to a kind of anorganic vitalism or panpsychism that sees sense everywhere yet fails to realize that life *makes sense* only to and for the living. Sense is not autonomous; it is *made* by living organisms that stir, feel, and act within their environments. For Deleuze, on the other hand, the virtual forces of becoming literally make sense on their own, and they do so without and beyond the stratified territory of living beings and human minds. This logic of sense is beyond bioaesthetics. If neo-Darwinian materialism is too reductionist, Deleuze and Guattari's spiritualism, in my view, is too speculative.

Consider, for example, the following passage from *What Is Philosophy?*: "The artist brings back from the chaos *varieties* that no longer constitute a reproduction of the sensory in the organ but set up a being of the sensory, a being of sensation, on an anorganic plane of composition that is able to restore the infinite."[23] One may easily agree with the first part of this statement: in neurological terms, neither our sense organs nor our brain takes a "picture" of how we actually "see" things according to our conventional codes and naturalized habits of perception. Rather, the incoming stimuli are mapped onto a neuronal network of chemical–electronic impulses. Since this neuronal map in no way resembles the kind of mental image humans consciously perceive in their minds, we might indeed call this unconscious (i.e., pre-phenomenological) neuronal map a "being of the sensory, a being of sensation."

But the second part of the statement remains puzzling nonetheless, because it reduces this map to an "anorganic plane of composition." Deleuze and Guattari effectively sever the "being of sensation" (i.e., the electrochemical impulse traveling along some neuronal network) from its physiological foundation in the sensing organism, which is being dissolved into an "anorganic plane of composition." What remains in the end is a message without a medium, a pure spirit that affects life itself. A look at Francis Bacon's paintings makes the case: "This ground, this rhythmic unity of the sense [portrayed by Bacon], can be discovered only by going beyond the organism. The phenomenological hypothesis is perhaps insufficient because it merely invokes the lived body. But the lived body is still a paltry thing in comparison with a more profound and almost unlivable Power."[24] My problem with this and similar claims by Deleuze is that the precise nature of this "almost unlivable Power" is impossible to conceptualize in either physiological or phenomenological terms without reterritorializing it onto the human body. Yet that is precisely what Deleuze's philosophy, more often than not, refuses to do. "Becoming," Deleuze famously claimed, "is not 'judged,' it is 'just' and possesses its own law in itself."[25] Deleuze's aesthetics is all about this "law in itself." It is wedded to the idea of life's passive (self-)transformation, which it pictures as a pure event that affects the human without being affected in return. His philosophy dematerializes living bodies to the point where they become nothing but "events of pure thought" and a matter of pure contemplation:[26]

> Sensation is pure contemplation, for it is through contemplation that one contracts, contemplating oneself to the extent that one contemplates the elements from which one originates. Contemplating is creating, the mystery of passive creation, sensation.[27]

This "mystery of passive creation" lies at the core of Deleuze's aesthetics. It is a myth similar to that of "Virgin Mary, Mother of God" used by Catholics to explain the immaculate conception of Jesus Christ. Much like Heidegger remained a Catholic philosopher throughout his life, Deleuze always remained a spiritualist at heart despite his materialism.

The same applies to Félix Guattari. Deleuze's juxtaposition of "the logic of sense" and common sense mirrors Guattari's juxtaposition of "a logic of affects" and rational discourse: "Affect is not a question of representation and discursivity," Guattari insists, "but of existence. I find myself transported into a Debussyst Universe, a blues universe, a blazing becoming of Provence. I have crossed a threshold of consistency."[28] Yet the question remains: who (or what) exactly is this "I" that allegedly undergoes these transportations into pure sensation? Deleuze and Guattari's answer, of course, is "life" itself, the virtual forces of eternal becoming. Deleuze certainly never made a secret of his fervent belief in vitalism: "Everything I've written is vitalistic, at least I hope it is."[29] But the being—or becoming—of life, I would insist, is not only rooted in some vital force, but also conceptualized as such in discourse. If we accept this bifurcated nature of life as both constructed and real, the epistemological quandary underneath the affective aesthetics of Deleuze and Guattari is easy to formulate. For how can any living organism—much less "I"—possibly make contact with the affective forces of pure becoming when the only way for us, humans, to experience these affects is precisely through our embodied modes of sensation and cognition? If that is so, does it not lead any and all aesthetics back to phenomenological experience and the history of the embodied mind?

To escape this destiny, Deleuze and Guattari's aesthetics moves beyond our *experience* of sensation toward us *becoming* sensation altogether. The best literary texts, in their view, are neither about meaning nor about being, but about becoming other. Hence Melville's Ahab is not just hunting the White Whale, but feels himself *becoming whale,* much like Lord Chandos in Hofmannsthal's *A Letter* is not just writing *about* rats, but writes about *himself becoming rat.* And if "writing is a becoming," so, too, is the act of reading.[30] Not only Gregor Samsa, but every Deleuzian reader of Kafka's *Metamorphosis* is becoming insect, becoming vermin. Art creates lines of flight for humans to flee their bodies and lose their minds. Francis Bacon thus "gives us eyes all over: in the ear, in the stomach, in the lungs," so that our eye "ceases to be organic in order to become a polyvalent and transitory organ" of pure becoming.[31]

It is far from self-evident how to read—and understand—such passages. Given Deleuze's explicit renunciation of metaphorical language, one has little choice but to read them literally. If so, then I would like to point out that there are many reasons why people cannot, all of a sudden, acquire eyes in the stomach: it would be difficult for them to see or to digest food. I do not mean to sound flippant, but there is little reason to assume that the aesthetic–affective transformations celebrated by Deleuze will be remotely as intriguing or rewarding an experience as he envisions it to be in front of Bacon's paintings. For all we know, our becoming animal might just as well result in the human mutations portrayed in Richard Matheson's novel *I Am Legend,* mindless creatures full of rage and hatred. Or we might end up like the nomads from the North described in one of Kafka's stories, tearing to pieces a living, breathing ox with our bare hands and teeth to fill our stomachs. Or we might look like any of the other countless zombies, vampires, and alien monsters that populate the sci-fi genre, TV programs, and the video game industry today. What else are *The Walking Dead* but walking affect? And why should we assume that our touching the stream of pure affect will result in a positive transformational experience for us, human beings? I might be overly pessimistic, but my fear is that we will end up like *The Fly* in David Cronenberg's film, whose most fervent wish is to be killed in order to end its misery of becoming animal.

We should also remember that according to Nietzsche—who, after all, had a profound influence on Deleuze's philosophy—aesthetic experience is not only meant to unleash, but also to shield us from Dionysian affect through Apollonian form. This, indeed, was the role of affect as defined in early eighteenth-century aesthetics as well. Prior to the rise of romanticism and its speculative theory of art, the sole reason to unleash the affective power of art was for the beholder to practice the *control* of affect, not its release. The goal was to learn to repress one's emotional responses to intense sensation in order to cultivate and adopt social conventions and public norms of behavior.[32] I have tried to show that biologism either implicitly or explicitly returns to this preromantic, prespeculative notion of art and aesthetics. Neo-Darwinian aesthetics uses art to reverse engineer our species' mental architecture allegedly shaped in the Pleistocene, while neuroaesthetics emphasizes the continuing power of art and aesthetics to recondition human perception and cognition.

Deleuze's posthuman aesthetics of affect approaches art from the opposite end of the spectrum. His goal is to deliver the human body *from* its habits not in order to create new ones, but in order to suspend them altogether.[33] Deleuze's aesthetics is an aesthetics of passivity and refusal. It is expressed most poignantly in the phrase "I would prefer not to" that Melville's "Bartleby, the Scrivener" utters whenever the world requires him to act. What matters most to Deleuze, in other words, is the hiatus, the interruption, the gap between stimulus and response, action and reaction. For these interruptions suspend the time of being. And this, precisely, opens up the space of haecceity, of self-transformation and pure becoming that founds the mystery of passive creation.

The suspension of time at the core of Deleuze's aesthetics is clearly related to the paradoxical temporal structure of Kant's aesthetic judgments I outlined in the first chapter, and I am heartened by scholars' recent efforts to examine and articulate this connection more clearly. John Protevi is one of several philosophers who acknowledge the importance of Kant for Deleuze's aesthetics. Yet he also argues that any rapprochement between the two must abandon the Kantian model of critical self-reflection that privileges reason, language, and intentionality. The transcendental conditions of sense making, Protevi reminds us, are enacted differently by different life forms, and he follows Deleuze in promoting a less cognitive and more autonomous sense of aesthetics that he calls a "new transcendental aesthetic" or an "evolutionary genetic phenomenology."[34] I certainly appreciate Protevi's argument and consider his body of work essential to bioaesthetics overall. Still, I cannot accept his (and Deleuze's) notion of larval subjectivity as anything but a vitalist metaphor that errs as much on the speculative side as neo-Darwinian theory errs on the reductionist side of how things are.

Steven Shaviro, too, has developed an intriguing conciliatory reading of Kant and Deleuze. He particularly praises Kant's notion of aesthetic judgment for entailing "a kind of sentience that is more than just a passive reception of sensible intuition but less than conscious recognition and comprehension."[35] Shaviro's attempt to carve out a space for aesthetic experience in between the passive and active mode of being strikes me as an excellent description of human creativity. It also reminds me of Derrida, who made a similar move when he attributed human thought and decision making to "the other in me." In one single gesture, Derrida both subjectified and deindividualized the process of human cognition, because the other in me is both "completely other" and yet uncannily familiar to me at the same time.[36] It follows that any decision I make remains ambiguously suspended between the active and passive mode of my being alive. "These are the paradoxes that are difficult to integrate in a classical philosophical discourse," Derrida conceded, "but I do not believe that a decision, if it exists, is possible otherwise."[37]

I think he is right, not least because we can interpret "the other in me" not only in religious–spiritual terms (i.e., the other as God or specter), but also in evolutionary–scientific terms (i.e., the other as species or ancestor). This second possibility is the one I would choose. Either way, bioaesthetics remains strongly committed to both cognitive science and evolutionary theory, a commitment more or less shared by Protevi and Shaviro as well. Hence the best way to distinguish bioaesthetics from both Protevi's "new transcendental aesthetic" and Shaviro's "speculative aesthetics," in my view, is to hold fast to Varela's idea that living is sense making. Like other speculative realists inspired by Deleuze and Whitehead, Shaviro does not always mind the categorical distinction between living organisms and inert matter, because his thought follows the body without organs rather than embodied modes of cognition. And like Protevi, Shaviro tends to overemphasize the passive nature of aesthetic experience as opposed to the decisive moment of aesthetic judgment—that is, the necessity of the human subject to enact itself *by making a decision.*

Bioaesthetics owes this insight not just to Kant, but to the rich history of interdisciplinary collaboration that extends from the Macy Conferences in the 1940s and 1950s to the rise of autopoiesis and general systems theory in the 1970s and beyond. Both cybernetics and systems theory emphasized that thinking is discriminatory by nature. It requires the drawing of distinctions, as Spencer-Brown famously said, because "a universe comes into being when a space is severed or broken apart."[38] Regardless of where exactly one wishes to draw the distinction (between me and the other or between a system and its environment), the crucial point is that we *cannot not* draw these distinctions, because "every living organism categorizes."[39] This it *must* do in order to come into being and to distinguish itself from its environment. Without this self-differentiation of organism and environment, life simply ceases to exist. No thought is comprehensive, and no life form is ubiquitous or without a center, not even a rhizome. Georges Canguilhem, the great historian of science and teacher of Foucault, emphasized again and again "that, in order for there to be an environment, there must be a center. It is the position of a living being, its relation to the experience it lives in as a totality, that gives the milieu meaning as conditions of existence. Only a living being, *infra*-human, can coordinate a milieu."[40] The activity of drawing distinctions creates the very reality we (humans) subsequently come to inhabit, and neither philosophy nor science makes sense outside this reality. Human knowledge only exists *because* of our embodied enactment of the world, not in spite of it. This realization of embodied cognition distinguishes bioaesthetics from biologism, and it distinguishes the sense-making activity of autopoietic systems from the autonomous sense found in Deleuze or larval subjects.

If we agree that "the mystery of passive creation" gives rise to an event, an encounter, an emergence, we should also agree that more often than not, such events are likely to implode on their own or vanish without a trace. The mere chance occurrence of genetic mutation, for example, means very little, as does a single synaptic event of neuronal connectivity in human brains. What matters is not only the occurrence of an event, but its ability to stay alive and affect its environment in a lasting way. What matters, in short, is the world that emerges *in the wake of an event.* Both in biological and cultural terms, the occurrence of an event constitutes the beginning, rather than the end, of the complex interactive process between organism and environment that characterizes life. The vast majority of the countless events that occur every millisecond in the universe, I would guess, will turn out to have been utterly insignificant in our world and without meaning for us, which is why I see little reason to think too much about them. What matters to bioaesthetics instead is the nature of being alive and the specificity of our experiences, which the humanities are called on to elucidate and make sense of—for us, humans.

NOTES

Introduction

1. Ian Hacking, *The Social Construction of What?* (Cambridge, Mass.: Harvard University Press, 1999), 3.
2. Nikolas Rose, for example, discusses the rise of a new economic system he calls "bioeconomy," a system based on the extraction of "biovalue" from "the vital properties of living processes." See Rose, *The Politics of Life Itself: Biomedicine, Power, and Subjectivity in the Twenty-First Century* (Princeton, N.J.: Princeton University Press, 2007), 32. Kaushik Sunder Rajan coins the term "biocapital" to denote "a form of enterprise inextricable from contemporary capitalism" and based on the "implosion of the economic and [the] epistemic." See Rajan, *Biocapital: The Constitution of Postgenomic Life* (Durham, N.C.: Duke University Press, 2006), 3, 283. Eugene Thacker's concept of "biomedia" makes the point that the "biological and the digital domains are no longer rendered ontologically distinct, but instead are seen to inhere in each other." See Thacker, *Biomedia* (Minneapolis: University of Minnesota Press, 2004), 7.
3. Clayton M. Christensen and Henry J. Eyring, *The Innovative University: Changing the DNA of Higher Education from the Inside Out* (Hoboken, N.J.: Jossey-Bass, 2011).
4. Joseph Carroll, *Reading Human Nature: Literary Darwinism in Theory and Practice* (Albany: State University of New York Press, 2011), 271.
5. Ellen Dissanayake, *Homo Aestheticus: Where Art Comes From and Why* (Seattle: University of Washington Press, 1992), xix.
6. Semir Zeki, *Inner Vision: An Exploration of Art and the Brain* (Oxford, U.K.: Oxford University Press, 1999), 217.
7. Friedrich Nietzsche, *Nachgelassene Fragmente*, 88–89, qtd. in Karlheinz Barck, "Differenzierungen im Ästhetikbegriff (Transformationen idealistischer Ästhetik)," in *Ästhetische Grundbegriffe*, ed. Karlheinz Barck et al., 7 vols. (Stuttgart: Metzler 2000–), 1:386; my translation.
8. Jean-Marie Schaeffer, *Art in the Modern Age: Philosophy of Art from Kant to Heidegger*, trans. Steven Rendall (Princeton, N.J.: Princeton University Press, 2000), 7.

9. Ibid.
10. Ellen Dissanayake, "The Arts after Darwin: Does Art Have an Origin and Adaptive Function?" in *World Art Studies: Exploring Concepts and Approaches,* ed. K. Zijlmans and W. van Damme (Amsterdam: Valiz, 2008), 244.
11. Cf. Eckart Voland and Karl Grammer, eds., *Evolutionary Aesthetics* (Berlin: Springer, 2003); Joseph Carroll, *Literary Darwinism: Evolution, Human Nature, and Literature* (New York: Routledge, 2004); Carroll, *Reading Human Nature*; Alan Richardson and Ellen Spolsky, eds., *The Work of Fiction: Cognition, Culture, and Complexity* (New York: Ashgate, 2004); Lisa Zunshine, ed., *Introduction to Cognitive Literary Studies* (Baltimore: Johns Hopkins University Press, 2010).
12. Brett Cooke, "Biopoetics: The New Synthesis," in *Biopoetics: Evolutionary Explorations in the Arts,* ed. Brett Cooke and Frederick Turner (Lexington, Ky.: ICUS, 1999), 23.
13. Carroll, *Reading Human Nature,* 271.
14. As Edward Slingerland recently put it, "There exist compelling reasons for thinking that empirical inquiry tells us something about the humanly meaningful world that armchair speculation does not." See Slingerland, *What Science Offers the Humanities: Integrating Body and Culture* (Cambridge, U.K.: Cambridge University Press, 2008), 248.
15. Cf. Geoffrey Harpham, *Language Alone: The Critical Fetish of Modernity* (New York: Routledge, 2002).
16. Jeffrey Nealon rightly points out that Deleuze's "move—or at least his *claim* to move—beyond signification and interpretation [was] the lynchpin of the warm reception that Deleuze has received in North America." See Nealon, "Beyond Hermeneutics," in *Between Deleuze and Derrida,* ed. Paul Patton and John Protevi (London: Continuum, 2003), 161.
17. Brian Massumi, *Parables of the Virtual: Movement, Affect, Sensation* (Durham, N.C.: Duke University Press, 2002), 1.
18. Hans-Georg Gadamer, *Truth and Method,* 2nd ed., trans. Joel Weinsheimer and Donald G. Marshall (Continuum: New York, 1999), 456.
19. Ibid., 572; my emphasis.
20. Ibid., 164.
21. Cf. Jennifer Washburn, *University, Inc.: The Corporate Corruption of Higher Education* (New York: Basic Books, 2006).
22. E. O. Wilson, *Consilience: The Unity of Knowledge* (New York: Knopf, 1998), 266.
23. Daniel Dennett, *Darwin's Dangerous Idea: Evolution and the Meanings of Life* (London: Penguin, 1995), 82.
24. Stephen Jay Gould, *The Hedgehog, the Fox, and the Magister's Pox: Mending the Gap between Science and the Humanities* (New York: Vintage Books, 2004), 5–6.
25. Elizabeth Grosz makes a similar point: "The natural sciences, even as they may be augmented by the social sciences, nevertheless remain unable to grasp the qualitative nuances that only the humanities—each in their different way—address. The humanities each address, without clear-cut borders, the human as a literary, linguistic, artistic, philosophical, historical, and culturally variable being. They re-

main irreplaceable to the extent that these questions have not been and perhaps cannot be addressed through other knowledges. However, they are not invariable." See Grosz, *Becoming Undone: Darwinian Reflections on Life, Politics, and Art* (Durham, N.C.: Duke University Press, 2011), 15.

26. In "From Max Scheler to Helmuth Plessner, ending with Arnold Gehlen," Roberto Esposito points out, "the *conditio humana* is literally constituted by the negativity that separates it from itself." Which is to say that the "alienation of man represents the indispensable condition of our own identity." See Esposito, *Bíos: Biopolitics and Philosophy,* trans. Timothy Campbell (Minneapolis: University of Minnesota Press, 2010), 48. Giorgio Agamben agrees: "Man has no specific identity other than the *ability* to recognize himself," because "*man is the animal that must recognize itself as human to be human*. . . . Homo sapiens, then, is neither a clearly defined species nor a substance; it is, rather, a machine or device for producing the recognition of the human." See Agamben, *The Open: Man and Animal,* trans. Kevin Attell (Stanford, Calif.: Stanford University Press, 2003), 26.
27. Wilson's quote continues as follows: "In the light of modern biology, what is then human nature? It is not the genes, which prescribe human nature. Nor is it just the universal traits of culture, such as the creation myths, incest taboos, and rites of passage, possessed by all societies. Rather, it is the inherited regularities of sensory and mental development that animate and channel the acquisition of culture." See Wilson, "Foreword from the Scientific Side," in *The Literary Animal: Evolution and the Nature of Narrative,* ed. Jonathan Gottschall and David Sloan Wilson (Evanston, Ill.: Northwestern University Press, 2005), viii. I will elaborate and critique this argument at length in chapters 3 and 4.
28. Eric Kandel, *Reductionism in Art and Brain Science: Bridging the Two Cultures* (New York: Columbia University Press, 2016), 187. By contrast, the cognitive neuroscientist Anjan Chatterjee gets it right: "It makes sense to try to understand the way art can be understood both biologically and culturally. Some questions will be better answered by biological methods and others through cultural analysis." Anjan Chatterey, *The Aesthetic Brain: How We Evolved to Enjoy Beauty and Enjoy Art* (Oxford, U.K.: Oxford University Press, 2015), 124.
29. Hannah Arendt, *The Human Condition* (Chicago: University of Chicago Press, 1998), 10–11.
30. Leda Cosmides, John Tooby, and Jerome H. Barkow, "Introduction: Evolutionary Psychology and Conceptual Integration," in *The Adapted Mind: Evolutionary Psychology and the Generation of Culture,* ed. Jerome H. Barkow, Leda Cosmides, and John Tooby (Oxford, U.K.: Oxford University Press, 1992), 6. William E. Connolly, too, points out that "if constructivism remains thin and untextured, and no *third way* is consolidated between it and the flat realism it challenges, the stage is set for a forceful return to the other side of a too-simple statement of alternatives." See Connolly, *Neuropolitics* (Minneapolis: University of Minnesota Press, 2002), 17; my emphasis. More recently, Ellen Spolsky, Lisa Zunshine, and Patrick Colm Hogan have each argued in favor of a "middle ground" that transcends the epistemological gap between the cognitivist and the literary-historical view of art. See

Zunshine, *Introduction to Cognitive Cultural Studies.* A notable exception is the philosopher Graham Harman, who shares my rejection of the allegedly "neutral" middle ground—using it, however, to support rather than question the "objectivity" of scientific realism: "To remain 'neutral' on this question [of scientific realism] is to condemn philosophy to operate only as a reflexive meta-critique of the condition of knowledge." That, indeed, is the best philosophy can do. See Harman, *Guerrilla Metaphysics: Phenomenology and the Carpentry of Things* (Chicago: Open Court, 2005), 42.

31. R. C. Lewontin, *Biology as Ideology: The Doctrine of DNA* (New York: Harper, 1993).
32. David Sloan Wilson, "Evolutionary Social Constructivism," in *Literary Animal,* ed. Gottschall and Wilson, 36. Similarly, see Ellen Spolsky, "Darwin and Derrida: Cognitive Literary Theory as a Species of Post-Structuralism," *Poetics Today* 23, no. 1 (2002): 43–62.
33. "'The environment' is as much a myth as is 'the gene.' Environments exist at multiple levels. Thus for an individual piece of DNA 'the environment' is all the rest of the DNA in the genome, plus the cellular metabolic system that surrounds it, proteins, enzymes, ions, water. . . . For organisms, the environment is constituted by the biological and physical world in which they move—and for humans, the social, cultural and technological world too." Steven Rose, *The Future of the Brain: The Promise and Perils of Tomorrow's Neuroscience* (Oxford, U.K.: Oxford University Press, 2005), 59.
34. Louis Menand, "Dangers Within and Without," in *MLA Profession, 2005,* ed. Rosemary G. Feal (New York: Modern Language Association Press, 2005), 14.
35. Cf. Ludwig Fleck, "Zur Krise der Wirklichkeit," *Die Naturwissenschaften* 17 (1925): 425–30; Gaston Bachelard, *The Formation of the Scientific Mind* (New York: Clinamen Press, 2006).
36. Hans-Jörg Rheinberger, *An Epistemology of the Concrete: Twentieth-Century Histories of Life* (Durham, N.C.: Duke University Press, 2010), 29.
37. Georges Canguilhem, *Knowledge of Life,* ed. Paola Marrati and Todd Meyers, trans. Stefanos Geroulanos and Daniela Ginsburg (New York: Fordham University Press, 2008), 22.
38. Lorraine Daston, "Science Studies and the History of Science," *Critical Inquiry* 35, no. 4 (2009): 813.
39. Ian Hacking, *Historical Ontology* (Cambridge, Mass.: Harvard University Press, 2002), 68.
40. Ian Hacking, *Representing and Intervening: Introductory Topics in the Philosophy of Natural Science* (Cambridge, U.K.: Cambridge University Press, 1983), 15.
41. Ibid., 14–15.
42. Hacking, *Historical Ontology,* 11.
43. Lorraine Daston and Peter Galison, *Objectivity* (New York: Zone Books, 2007), 16.
44. Peter Godfrey-Smith's advice is to "avoid the term [objectivity], because it is ambiguous and tends to set up the issues in a misleading way. . . . It is crude, and it tends to suggest false dichotomies." See Godfrey-Smith, *Theory and Reality: An*

Introduction to the Philosophy of Science (Chicago: University of Chicago Press, 2003), 228–29.

45. According to the *OED*, objectivity is "the ability to consider or represent facts, information, etc., without being influenced by personal feelings or opinions" (see http://www.oed.com/view/Entry/129639?redirectedFrom=objectivity#). The pragmatist philosopher Richard Rorty defines objectivity as a "property of theories which, having been thoroughly discussed, are chosen by a consensus of rational discussants." See Rorty, *Philosophy and the Mirror of Nature* (Princeton, N.J.: Princeton University Press, 1979), 338. By contrast, the historian of science Olaf Breidbach laconically states, "Objectivity, then, is a subjective construct." See Breidbach, "The Beauties and the Beautiful: Some Considerations from the Perspective of Neuronal Aesthetics," in *Evolutionary Aesthetics*, ed. Voland and Grammer, 56. The cyberneticist Heinz von Förster, finally, "humbly declares that the claim for objectivity is nonsensical *[unsinnig]*." See von Förster, *KybernEthik*, trans. Birger Ollrogge (Berlin: Merve, 1993), 88; my translation.
46. Gregory Bateson, *Steps to an Ecology of Mind: Collected Essays in Anthropology, Psychiatry, Evolution, and Epistemology* (Chicago: University of Chicago Press, 2000), 47.
47. Boris Groys, *The Communist Postscript*, trans. Thomas Ford (London: Verso, 2009), 7.
48. Eugene Thacker, *After Life* (Chicago: University of Chicago Press, 2010), xiii.
49. Hacking, *Historical Ontology*, 64.
50. Gianni Vattimo, *Beyond Interpretation: The Meaning of Hermeneutics for Philosophy*, trans. David Webb (Stanford, Calif.: Stanford University Press, 1997), 107. See also Michel Foucault, "What Is an Author?" in *The Norton Anthology of Theory and Criticism*, ed. Vincent B. Leitch et al., 2nd ed. (New York: Norton, 2010), 1475–90.
51. Rorty, *Philosophy and the Mirror of Nature*, 320. I should point out that Rorty opposes hermeneutics to epistemology (cf. 318), whereas I oppose hermeneutics to science and argue that both, hermeneutics and science, are epistemological systems. Regardless of this terminological difference, however, the substance of the argument remains the same, which is that hermeneutics cannot be reduced to scientific inquiry, but yields knowledge of a different kind.
52. "The hermeneutic phenomenon is basically not a problem of method at all," Gadamer repeats again and again, because "understanding is not just one of the various possible behaviors of the subject but the mode of being of Dasein itself." See Gadamer, *Truth and Method*, xxi–xxii, xxx. On this crucial methodological point, at least, Jacques Derrida agrees with his lifelong adversary Gadamer: "All sentences of the type 'deconstruction is X' or 'deconstruction is not X' *a priori* miss the point . . . [which] is precisely the delimiting of ontology and above all of the third person indicative: S *is* P." See Derrida, "Letter to a Japanese Friend," in *A Derrida Reader: Between the Blinds*, ed. Peggy Kamuf (New York: Columbia University Press, 1991), 275.
53. Gadamer, *Truth and Method*, xxiii.

54. Barck et al., "Vorwort," in *Ästhetische Grundbegriffe,* ed. Barck et al., 1:x; my translation.

55. Most scientists, Lorraine Daston contends, "take it for granted that things are simultaneously material and meaningful." See Daston, "Introduction," in *Things That Talk: Object Lessons from Art and Science,* ed. Lorraine Daston (New York: Zone Books, 2004), 17. That has been my personal experience in working with scientists as well.

56. Bruno Latour, *Pandora's Hope: Essays on the Reality of Science Studies* (Cambridge, Mass.: Harvard University Press, 1999), 127.

57. There are two radically different definitions of the "gene" concept that continue to exist side by side. In its original, classical sense introduced by Gregor Mendel and Thomas Hunt Morgan, a gene is an abstract construct that points to an observable trait. In his acceptance speech from 1933, Morgan still emphasized that there "is no consensus of opinion amongst geneticists as to what genes are—whether they are real or purely fictitious," because this distinction makes no difference "at the level at which genetic experiments lie." Morgan, qtd. in Michael R. Hendrickson, "Exorcising Schrödinger's Ghost: Reflections on 'What Is Life?' and Its Surprising Relevance to Cancer Biology," in *What Is Life? The Intellectual Pertinence of Erwin Schrödinger,* by Hans Ulrich Gumbrecht, Robert Harrison, Michael R. Hendrickson, and Robert B. Laughlin (Stanford, Calif.: Stanford University Press, 2011), 55. The second sense of gene is more recent and defines it as a segment of DNA that codes for mRNA.

58. Thierry Bardini, *Junkware* (Minneapolis: University of Minnesota Press, 2011), 50.

59. Gilles Deleuze, *Difference and Repetition,* trans. Paul Patton (New York: Columbia University Press, 1994), 78.

60. John Protevi, *Life, War, Earth: Deleuze and the Sciences* (Minneapolis: University of Minnesota Press, 2013), 175.

61. Protevi provides a comprehensive account of protein production in *Life, War, Earth.* See ibid., 198ff.

62. Evan Thompson, *Mind in Life: Biology, Phenomenology, and the Sciences of Mind* (Cambridge, Mass.: Harvard University Press, 2007), 57.

63. Michael Ruse, "Introduction," in *The Oxford Handbook of Philosophy of Biology,* ed. Michael Ruse (Oxford, U.K.: Oxford University Press, 2008), 10.

64. Thomas Lemke, *Biopolitics: An Advanced Introduction,* trans. Eric Frederick Trump (New York: New York University Press, 2011), xi. Roberto Esposito agrees: "Compressed (and at the same time destabilized) by competing readings and subject to continuous rotations of meaning around its own axis, the concept of biopolitics risks losing its identity and becoming an enigma." See Esposito, *Bíos,* 15.

65. Carl Sagan, "Life," qtd. in Ed Regis, *What Is Life? Investigating the Nature of Life in the Age of Synthetic Biology* (Oxford, U.K.: Oxford University Press, 2008), 156. "In the end," the mathematician Stephen Wolfram recently concluded, "every single general definition that has been given both includes systems that are not normally considered alive, and excludes ones that are. (Self-reproduction, for example, suggests that flames are alive, but mules are not.)" See Wolfram, *A New Kind of Science*

(Champaign, Ill.: Wolfram Media, 2002), 1178. Édouard Machery makes the same point: "There is no way to decide between heterogeneous definitions of life" (qtd. in Regis, *What Is Life?* 159). "Life will self-transcend," Lynn Margulis and Dorion Sagan conclude, "any definition slips away." See Margulis and Sagan, *What Is Life?* (Berkeley: University of California Press, 2000), 217–18.

66. Erwin Schrödinger, "What Is Life? The Physical Aspect of the Living Cell," in *What Is Life? With Mind and Matter and Autobiographical Sketches* (Cambridge, U.K.: Cambridge University Press, 1967), 1–90.
67. Ibid., 5.
68. "Living matter, while not eluding the 'laws of physics' as established up to date, is likely to involve 'other laws of physics' hitherto unknown, which, however, once they have been revealed, will form just as integral a part of this science as the former" (ibid., 68).
69. It should be noted that the leading physicists of the twentieth century (like Werner Heisenberg and Niels Bohr) were much more cautious in their assessment of science: "Indeed, the essential non-analyzability of atomic stability in mechanical terms presents a close analogy to the impossibility of a physical or chemical explanation of the peculiar functions characteristic of life." See Bohr, *Atomic Physics and Human Knowledge* (1961; repr., Mineola, N.Y.: Dover, 2010), 9.
70. Schrödinger, *What Is Life?* 22.
71. J. D. Watson and F. H. C. Crick, "Molecular Structure of Nucleic Acids: A Structure for Deoxyribose Nucleic Acid," *Nature* 171, no. 25 (1953): 737–40.
72. Hendrickson, "Exorcising Schrödinger's Ghost," 100. According to Hendrickson, this fundamental "conceptual change in biology" from the classical dogma of molecular biology to the Post-Schrödingerian Perspective can be summarized in the following theses: first, "DNA, unaided, is not self-replicating"; second, "the hereditary stability of DNA is an emergent property"; third, "the genome is not the sole hereditary resource," fourth, "the genome (totality of DNA) is structurally dynamic" (68ff.). "In summary," Hendrickson states, "the genomic DNA increasingly appears to be more of a database of genetic resources than Schrödinger's all-powerful director" (71). Evelyn Fox Keller agrees: "Contrary to all expectations," she writes, "instead of lending support to the familiar notions of genetic determinism that have acquired so powerful a grip on the popular imagination, these successes [of the genome project] pose critical challenges to such notions." Keller, *The Century of the Gene* (Cambridge, Mass.: Harvard University Press, 2000), 5.
73. Margulis and Sagan, *What Is Life?* 14.
74. Tellingly titled *The Poetics of DNA*, Judith Roof's study denounced much of contemporary genetic theory as "pseudoscientific ideas that bypass scientific reasoning in favor of magic, instantaneity, and the commodity short-circuiting of work and desire" (3). See Roof, *The Poetics of DNA* (Minneapolis: University of Minnesota Press, 2007). See also Evelyn Fox Keller, *Making Sense of Life: Explaining Biological Development with Models, Metaphors, and Machines* (Cambridge, Mass.: Harvard University Press, 2002); Richard Lewontin, *It Ain't Necessarily So: The Dream of the Human Genome and Other Illusions* (New York: New York Review of

Books, 2000); Keller, *Century of the Gene*; Lily E. Kay, *Who Wrote the Book of Life? A History of the Genetic Code* (Stanford, Calif.: Stanford University Press, 2000); Ruth Hubbard and Elijah Wald, *Exploding the Gene Myth: How Genetic Information Is Produced and Manipulated by Scientists, Physicians, Employers, Insurance Companies, Educators, and Law Enforcers* (Boston: Beacon, 1997).

75. Hacking, *Historical Ontology*, 23.
76. Evelyn Fox Keller, *The Mirage of a Space between Nature and Nurture* (Durham, N.C.: Duke University Press, 2010), 34, 9, 10. As she puts it elsewhere in the book, "At least part of the explanation of the unreasonable persistence of the [nature/nurture] debate is to be found in the language of particulate genes, a language that originally developed out of the hope that a particulate theory of inheritance might do for biology what the atomic theory had done for chemistry. For good or bad, biology has turned out to be vastly more complicated, and so has genetics itself" (13).
77. Ruse, "Introduction," 3.
78. "There are not only the manifolds of space and time in which things can be spread out. There is also the manifold of environments, in which things repeat themselves in always new forms." Jakob von Uexküll, *A Foray into the Worlds of Animals and Humans*, with *A Theory of Meaning*, trans. Joseph D. O'Neil (Minneapolis: University of Minnesota Press, 2010), 208.
79. Cf. Terrence W. Deacon, *Incomplete Nature: How Mind Emerged from Matter* (New York: Norton, 2011).
80. Xavier Bichat, qtd. in Canguilhem, *Knowledge of Life*, 69.
81. Canguilhem, *Knowledge of Life*, 71.
82. "A reduced biology has as its corollary the effacement of the biological object as such—in other words, the devaluation of its specificity" (ibid., 59). Ernst Mayr, too, points out that "many of the generalizations derived from such philosophies of physics are irrelevant when applied to biology." And the reverse is true as well—namely, that "many phenomena and findings of the biological sciences have no equivalent in the physical sciences and are therefore omitted from philosophies of science that are based on physics." See Mayr, "Discussion: Footnotes on the Philosophy of Biology," *Philosophy of Science* 36, no. 2 (1949): 198. W. W. Bartley goes even further and claims that biological theory not only differs, but actually "conflicts with" the philosophy of science based on physics. See Gerard Radnitzky and W. W. Bartley III, eds., *Evolutionary Epistemology, Rationality, and the Sociology of Knowledge* (La Salle, Ill.: Open Court, 1987), 8.
83. Steven Rose, *The 21st-Century Brain: Explaining, Mending and Manipulating the Mind* (London: Vintage Books, 2005), 57.
84. Barck et al., "Vorwort," in *Ästhetische Grundbegriffe*, ed. Barck et al., 1:vii; my translation. In their preface, the editors provide a comprehensive list of pertinent examples that include "formulations such as 'Aesthetics of Existence' (Michel Foucault), 'Aesthetics of Disappearing' (Paul Virilio), 'Ethic of Aesthetics' (Michel Maffesoli), 'Aesthetics of Fright' (Karl Heinz Bohrer), 'Aesthetics of Violence' (Wolf Lepenies) and even 'Aesthetics of Science' (Judith Wechsler)." See ibid., 1:x; my translation.

85. Ibid., 1:311; my translation.
86. Ibid., 1:392; my translation.
87. For more details, see Keller, *Making Sense of Life.*
88. Rachel Zuckert, *Kant on Beauty and Biology: An Interpretation of the Critique of Judgment* (Cambridge, U.K.: Cambridge University Press, 2007), 17.
89. Dennett, *Darwin's Dangerous Idea,* 144.
90. See Claire Colebrook, *Death of the PostHuman: Essays on Extinction, Vol. 1* (New York: Open Humanities Press, 2014). As Brian Massumi puts it, "techniques of negative critique [must] be used sparingly" in order "to shift to [more] affirmative methods" of investigation. See Massumi, *Parables of the Virtual,* 13.

1. Human Nature after Kant

1. Cf. C. P. Snow, *The Two Cultures* (Cambridge, U.K.: Cambridge University Press, 1998).
2. Immanuel Kant, *Critique of Pure Reason,* ed. and trans. Marcus Weigelt (New York: Penguin, 2007), 60. All translations of Kant's *First Critique* refer to this edition (henceforth abbreviated *CPR*) unless noted otherwise. References to the German edition are from Immanuel Kant, *Kritik der Reinen Vernunft,* in *Werke in Sechs Bänden,* ed. Wilhelm Weischedel, vol. 2 (Darmstadt: Wissenschaftliche Buchgesellschaft, 1957) (henceforth abbreviated *KdRV*), 70, A21/B36. The page numbers after A and B refer to the first and second German editions, respectively.
3. Ibid. For more details, see Howard Caygill, *A Kant Dictionary* (Hoboken, N.J.: Wiley-Blackwell, 1995), 53.
4. Cf. Peter Gilgen, "Aesthetics," in *The Blackwell Encyclopedia of Literary and Cultural Theory,* 3 vols., ed. Michael Ryan et al. (Oxford, U.K.: Blackwell, 2011), 122–28.
5. Immanuel Kant, "Beweisgrund zu einer Demonstration des Daseins Gottes," qtd. in Phillip R. Sloan, "Performing the Categories: Eighteenth-Century Generation Theory and the Biological Roots of Kant's A Priori," *Journal of the History of Philosophy* 40, no. 2 (2002): 237.
6. Kant, *CPR,* 653; *KdRV,* 696, A833/B861.
7. Kant, *CPR,* 610 (translation altered); *KdRV,* 649, A766/B794.
8. Kant, *CPR,* 95; *KdRV,* 108, A66/B91.
9. Kant, *CPR,* 168; *KdRV,* 158, B167.
10. Depending on which source one chooses to consult on the subject, the honor of having coined the term "biology" to designate a new scientific discipline goes to Gottfried Reinhold Treviranus for his *Biologie, oder die Philosophie der lebenden Natur* or to Jean-Baptiste Lamarck for his *Hydrogéologie,* both published in 1802 in Germany and France, respectively.
11. Kant refers to chemistry and the life sciences as "improper sciences" *(uneigentliche Wissenschaften)*; see his preface to the *Metaphysische Anfangsgründe der Naturwissenschaft,* in *Gesammelte Schriften, Akademieausgabe,* vol. 4 (Berlin: De Gruyter, 1969), 468.
12. Kant, *CJ,* 271; *KdU,* 515, A333–34/B337–38.

13. Kant, *CJ*, 292; *KdU*, 545, A374/B378–79.
14. Jennifer Mensch, *Kant's Organicism: Epigenesis and the Development of Critical Philosophy* (Chicago: University of Chicago Press, 2013), 15.
15. Humberto R. Maturana and Francisco J. Varela, *The Tree of Knowledge: The Biological Roots of Human Understanding*, trans. Robert Paolucci, 2nd ed. (Boulder, Colo.: Shambhala, 1992), 46–47. See also Humberto R. Maturana and Francisco J. Varela, *Autopoiesis and Cognition: The Realization of the Living* (Dordrecht: D. Reidel, 1980): "An autopoietic machine is a machine organized (defined as a unity) as a network of processes of production (transformation and destruction) of components which: (i) through their interactions and transformations continuously regenerate and realize the network of processes (relations) that produced them; and (ii) constitute it (the machine) as a concrete unity in space in which they (the components) exist by specifying the topological domain of its realization as such a network" (78).
16. Maturana and Varela, *Tree of Knowledge*, 244.
17. As Evan Thompson and others point out, there remains some disagreement among philosophers and scientists today about whether autopoietic systems are also cognitive systems. See Evan Thompson, *Mind in Life: Biology, Phenomenology, and the Sciences of Mind* (Cambridge, Mass.: Harvard University Press, 2007), 124ff. I shall further comment on the relation between autopoiesis and cognition later in this chapter and in chapter 2.
18. Although I have consulted a broad variety of primary sources, my account of the complex debate between preformationism and epigenesis in the eighteenth century relies a good deal on historical overviews and analyses provided by historians of science, including J. Wubnig, "The Epigenesis of Pure Reason," *Kant-Studien* 60, no. 2 (1969): 147–52; A. C. Genova, "Kant's Epigenesis of Pure Reason," *Kant-Studien* 65, no. 3 (1974): 259–73; Timothy Lenoir, "Kant, Blumenbach, and Vital Materialism in German Biology," *Isis* 71, no. 1 (1980): 77–108; James L. Larson, *Interpreting Nature: The Science of Living Form from Linnaeus to Kant* (Baltimore: Johns Hopkins University Press, 1994); Jacques Roger, *The Life Sciences in Eighteenth-Century French Thought*, ed. Keith Benson, trans. Robert Ellrich (Stanford, Calif.: Stanford University Press, 1997); Helmut Müller-Sievers, *Self-Generation: Biology, Philosophy, and Literature around 1800* (Stanford, Calif.: Stanford University Press, 1997); Robert J. Richards, "Kant and Blumenbach on the *Bildungstrieb*: A Historical Misunderstanding," *Studies in the History and Philosophy of Biology and Biomedical Sciences* 31, no. 1 (2000): 11–32; Sloan, "Performing the Categories"; John H. Zammito, "Kant's Persistent Ambivalence towards Epigenesis, 1764–90," in *Understanding Purpose: Kant and the Philosophy of Biology*, ed. Philippe Huneman (Rochester, N.Y.: University of Rochester Press, 2007), 51–74; Jocelyn Holland, *German Romanticism and Science: The Procreative Poetics of Goethe, Novalis, and Ritter* (London: Routledge, 2012); Mensch, *Kant's Organicism*. The following remarks are informed by and draw from these and other secondary literature as noted in the text.

19. The distinctive features of strong preformationism were first articulated by Jacques Roger in *The Life Sciences in Eighteenth-Century French Thought.*
20. Cf. Mensch, *Kant's Organicism,* 178n114.
21. Georges Buffon, qtd. in ibid., 41.
22. Cf. Mensch, *Kant's Organicism,* 48.
23. "In the same mode as gravity penetrates all parts of matter, so the power which impels or attracts the organic particles to food, penetrates into the internal parts of organized bodies, and as those bodies have a certain form, which we call the internal mould, the organic particles, impelled by the action of the penetrating force, cannot enter therein but in a certain order relative to this form, which consequently it cannot change, but only augment its dimension, and thus produce the growth of organized bodies." Buffon, qtd. in ibid., 42.
24. Buffon, qtd. in Larson, *Interpreting Nature,* 138.
25. As Larson states, "By adding penetrating forces that were analogous to attraction but operative only on organic matter, Buffon rendered the process of organic formation entirely unintelligible." Larson, *Interpreting Nature,* 137.
26. Haller, qtd. in ibid., 140.
27. Wolff, qtd. in ibid., 163.
28. Haller, qtd. in ibid., 140.
29. Haller, qtd. in Sloan, "Performing the Categories," 236.
30. Cf. Kant, "Von den verschiedenen Rassen der Menschen," qtd. in Sloan, "Performing the Categories," 240: "The foundations *[Gründe]* which lie in the nature of an organic body (plant or animal) for a determinate unfolding *[bestimmten Auswicklung]* are called germs *[Keime]* when this unfolding affects specific parts. But when it affects only the size of the relations of the parts to one another, I call them natural predispositions *[natürliche Anlagen].*"
31. Ibid.
32. Kant, *CPR,* 95; *KdRV,* 108, A66/B91.
33. Kant, "Von den verschiedenen Rassen der Menschen," in *Werke in Sechs Bänden,* vol. 4, 17, A6/B139; my translation.
34. Ibid.; my translation.
35. Kant, "Bestimmung des Begriffs einer Menschenrasse," in *Werke in Sechs Bänden,* vol. 4, 73, A403.
36. Ibid., 75, A406.
37. Ibid., 82, A417.
38. Kant, "Von den verschiedenen Rassen der Menschen," in *Werke in Sechs Bänden,* vol. 4, 28, B159.
39. Zammito, "Kant's Persistent Ambivalence toward Epigenetics," 1764.
40. Kant, *CJ,* 245; *KdU,* 486, A288/B292.
41. Ibid., *CJ,* 244; *KdU,* 484, A285/B289.
42. Blumenbach, qtd. in Richards, "Kant and Blumenbach on the *Bildungstrieb,*" 18.
43. Ibid., 25.
44. Wolff, qtd. in ibid., 15.

45. Cf. Larson, *Interpreting Nature*, 163ff.; and Timothy Lenoir, *The Strategies of Life: Teleology and Mechanics in Nineteenth-Century German Biology* (Chicago: University of Chicago Press, 1989), 21.
46. Wolff, qtd. in Larson, *Interpreting Nature*, 164.
47. Kant, "Von den verschiedenen Rassen der Menschen," qtd. in Sloan, "Performing the Categories," 244.
48. Blumenbach, qtd. in Richards, "Kant and Blumenbach on the *Bildungstrieb*," 24.
49. It remains unclear how familiar Kant was with the specifics of Wolff's theory. Philippe Huneman in particular emphasizes Kant's indebtedness to Wolff. Cf. Philippe Huneman, "Reflective Judgment and Wolffian Embryology: Kant's Shift between the First and the Third Critiques," in *Understanding Purpose*, ed. Huneman, 88. According to Sloan, however, there is "no evidence that Kant ever read any of Wolff's works directly," although there is evidence that he had indirect access to Wolff's ideas through other writers like Teten and Herder. Sloan, "Performing the Categories," 242n48.
50. Kant, qtd. in Lenoir, "Kant, Blumenbach, and Vital Materialism," 79–80.
51. Kant, *CJ*, 292–93; *KdU*, 545, A374/B378–79.
52. Richards, "Kant and Blumenbach on the *Bildungstrieb*," 25–26.
53. Kant, *CJ*, 245–46; *KdU*, 485–86, A287-89/B291–93.
54. Friedrich Nietzsche, "*Jenseits von Gut und Böse*," in *Werke in Fünf Bänden*, ed. Karl Schlechta (Frankfurt: Ullstein, 1981), vol. 3, 576; my translation.
55. Blumenbach, qtd. in Richards, "Kant and Blumenbach on the *Bildungstrieb*," 24.
56. Helmut Müller-Sievers develops this point at length in his study *Self-Generation: Biology, Philosophy, and Literature around 1800*.
57. Philippe Huneman, "Introduction," in *Understanding Purpose*, ed. Huneman, 16.
58. Cf. Lenoir, *Strategies of Life*.
59. For a good critique of Lenoir, see Richards, "Kant and Blumenbach on the *Bildungstrieb*." Less convincing, in my view, is Kenneth Caneva's convoluted book review of Lenoir's *Strategy of Life* in "Teleology with Regrets," *Annals of Science* 47 (1990): 291–300.
60. Gerhard Vollmer, *Biophilosophie* (Stuttgart: Reclam, 1995), 120; my translation.
61. Jacques Monod, *Chance and Necessity: An Essay on the Natural Philosophy of Modern Biology*, trans. Austryn Wainhouse (New York: Knopf, 1971), 154.
62. Cary Wolfe, *Animal Rites: American Culture, the Discourse of Species, and Posthumanist Theory* (Chicago: University of Chicago Press, 2003), 1–2.
63. "The resemblance between Kant's conception of self-organization and the current scientific one is striking," because "Kant recognized and formulated, probably for the first time, the autonomous organization proper to living beings." See Thompson, *Mind in Life*, 136, 211. Terrence Deacon agrees: "Probably the most prescient and abstract characterization of the dynamic logic of organism design was provided by the philosopher Immanuel Kant. . . . Although modern accounts can be far more concrete and explicit than Kant's, by virtue of their incorporation of over two hundred years of biological science, this knowledge can also be a source of distraction. . . . Today, it is possible to add flesh to Kant's skeletal definition and in

so doing demonstrate its prescience." See Terrence W. Deacon, *Incomplete Nature: How Mind Emerged from Matter* (New York: Norton, 2012), 302.

64. Thompson, *Mind in Life,* 149.
65. Ibid., 125.
66. See Protevi, *Life, War, Earth,* 17ff.
67. Deacon, *Incomplete Nature,* 305, 309.
68. Thompson, *Mind in Life,* 123.
69. Deacon, *Incomplete Nature,* 295.
70. Cf. Thompson, *Mind in Life,* 127.
71. Margulis and Sagan, *What Is Life?* 122.
72. Cf. Thompson, *Mind in Life,* 161. I return to the question of consciousness in my fifth chapter, on neuroaesthetics.
73. Deacon, *Incomplete Nature,* 309, 311, 509, 180.
74. Deacon stresses again and again that "*real teleological and intentional phenomena can emerge from physical and chemical processes previously devoid of these properties*" (ibid., 323).
75. Ibid., 315.
76. Ibid., 291.
77. Thompson, *Mind in Life,* 167. Margulis and Sagan agree: "All living beings," they argue, "evolved from the ancient common ancestor which evolved autopoiesis and thus became the first living cell" (*What Is Life?,* 48).
78. Terrence Deacon, "The Aesthetic Faculty," in *The Artful Mind: Cognitive Science and the Riddle of Human Creativity,* ed. Mark Turner (Oxford, U.K.: Oxford University Press, 2006), 23.
79. Varela, qtd. in Thompson, *Mind in Life,* 158.
80. Francisco J. Varela, Evan Thompson, and Eleanor Rosch, *The Embodied Mind* (Cambridge, Mass.: MIT Press, 1991), 202, 205.
81. Margulis and Sagan, *What Is Life?,* 14.
82. Francis Steen, "A Cognitive Account of Aesthetics," in *Artful Mind,* ed., Turner, 61, 70.
83. I should note that some neuroscientific studies have provided empirical evidence for this relation between our sense of identity and aesthetic experience, even though the data is limited and its interpretation fairly speculative at this point. Cf. Kandel, *Reductionism in Art and Brain Science,* 184.
84. Kant, *CJ,* 102. "Untersuchung der Frage: ob im Geschmacksurteile das Gefühl der Lust vor der Beurteilung des Gegenstandes, oder diese vor jener vorhergehe." See Kant, *KdU,* 295, A28/B28.
85. Ibid. "Die Auflösung dieser Frage ist der Schlüssel zur Kritik des Geschmacks und daher aller Aufmerksamkeit würdig."
86. Ibid. "Also ist es die allgemeine Mittheilungsfähigkeit des Gemüthszustandes in der gegebenen Vorstellung, welche, als subjective Bedingung des Geschmacksurtheils, demselben zum Grunde liegen, und die Lust am Gegenstand zur Folge haben muß."
87. Cf. Paul Guyer, *Kant and the Claims of Taste* (Cambridge, U.K.: Cambridge

University Press, 1979), 166; qtd. in Peter Gilgen, "Der Schlüssel zur Kritik des Geschmacks," in *Beobachtung Zweiter Ordnung im Historischen Kontext: Niklas Luhmann in Amerika*, ed. Perla Chinchilla Pawling, Aldo Mazzuchelli, Hans Ulrich Gumbrecht (Munich: Fink, 2013), 74.

88. Gilgen, "Der Schlüssel zur Kritik des Geschmacks," 91; my translation.
89. Cf. Niklas Luhmann, *Die Wissenschaft der Gesellschaft* (Frankfurt: Suhrkamp, 1992), 68–121.
90. Cf. Kant, *CJ*, 66–67; *KdU*, 251, Axxiv/Bxxvi.
91. Kant, *CJ*, 121. "Sondern sie kann als Notwendigkeit, die in einem ästhetischen Urteile gedacht wird, nur exemplarisch genannt werden, d.i. eine Notwendigkeit der Beistimmung aller zu einem Urteil, was wie Beispiel einer allgemeinen Regel, die man nicht angeben kann, angesehen wird." See Kant, *KdU*, 320, A63/B64.
92. Throughout the *Third Critique*, Kant equivocates between "judgments of taste" and "aesthetic judgments," even though part of his argument rests on the distinction between "mere judgments of taste" *(bloßes Geschmacksurteil)*—or what Kant also calls "mere sensual judgments" *(bloßes Sinnesurteil)*—on the one hand, and genuinely "reflective" (aesthetic) judgments *(reflektierende Urteilskraft)*, on the other. The former (i.e., *bloßes Geschmacksurteil/Sinnenurteil*) qualify objects that are "merely agreeable" *(bloß angenehm)*, whereas the latter qualify objects that are truly beautiful. To avoid confusion, I shall use the term "aesthetic judgment" to refer only to the latter case.
93. Cf. J. Grimm and W. Grimm, *Deutsches Wörterbuch von Jacob Grimm und Wilhelm Grimm*, vol. 26 (1854; repr., Munich: dtv, 1999), 1679.
94. Ibid., 1684; my translation, my emphasis.
95. Cf. Donald E. Brown, "The Universal People," in *Evolution, Literature, and Film: A Reader*, ed. Brian Boyd, Joseph Carroll, and Jonathan Gottschall (New York: Columbia University Press, 2010), 83–95.
96. Kant, *CJ*, 122. "Also müssen sie [Geschmacksurteile] ein subjectives Prinzip haben, welches nur durch Gefühl und nicht durch Begriffe, doch aber allgemeingültig bestimme, was gefalle oder missfalle." See *KdU*, 321, A64/B65.
97. Cf. *CJ*, 174; *KdU*, 391, A158/B160. Kant's central point is "that taste can be called *sensus communis* with greater justice than can the healthy understanding, and that the aesthetic power of judgment rather than the intellectual can bear the name of a communal sense." To retain this terminological distinction, which is crucial to my argument, I shall use "common sense" to denote the British, intellectual notion of communal sense, and *sensus communis* to identify Kant's notion of the *sensus communis aestheticus*.
98. Kant, *CJ*, 124; translation altered. I have replaced "common sense" with *sensus communis* in Guyer's otherwise excellent translation, because Kant's use of *Gemeinsinn* in this passage clarifies that he is talking about the *sensus communis aestheticus* and not the *sensus communis logicus*. See Kant, *KdU*, 323, A67/B68: "Diese unbestimmte Norm eines Gemeinsinns wird von uns wirklich vorausgesetzt: das beweiset unsere Anmaßung, Geschmacksurteile zu fällen."
99. Ibid.; translation altered, same as above. "Ob es in der Tat einen solchen Gemein-

sinn, als konstitutives Prinzip der Möglichkeit der Erfahrung gebe, oder ein noch höheres Prinzip der Vernunft es uns nur zum regulativen Prinzip mache, allererst einen Gemeinsinn zu höheren Zwecken in uns hervorzubringen; ob also Geschmack ein ursprüngliches und natürliches, oder nur die Idee von einem noch zu erwerbenden und künstlichen Vermögen, so daß ein Geschmacksurteil, mit seiner Zumutung einer allgemeinen Beistimmung, in der Tat nur eine Vernunftforderung sei, eine solche Einhelligkeit und Sinnesart hervorzubringen, . . . : das wollen und können wir hier noch nicht untersuchen." See Kant, *KdU,* 323–24, A67/B68.

100. Kant, *CJ,* 123; *KdU,* A67/B68.
101. Wilson, "Foreword from the Scientific Side," viii.

2. Marxism and Biology

1. Peter Singer, *A Darwinian Left: Politics, Evolution, and Cooperation* (New Haven, Conn.: Yale University Press, 2000), 5.
2. Margulis and Sagan suggest that cooperation might even be more fundamental to the emergence of life than competition. Margulis and Sagan, *What Is Life?,* 214.
3. Singer, *Darwinian Left,* 24.
4. Ibid., 37.
5. Ibid., 40, 31.
6. David Stack is foremost among those who seek to "forestall a growing movement to use Darwinism as the foundation for a new politics of the left," which he considers a "folly . . . course of action" (vii). Stack strongly contests Singer's claim that the left ignored evolutionary theory, arguing instead that "Darwinism was woven into the pattern of late-nineteenth- and early-twentieth-century socialism" (3). In his detailed critique, Stack repudiates Singer's account on philosophical, political, and historical grounds. Philosophically, he rightly points out that Singer "confounds evolutionary psychology with Darwinism" (4) and fails to take seriously the numerous critics of evolutionary psychology (e.g., Steven Rose, Stephen Jay Gould, Richard Lewontin). Politically, Singer's embrace of evolutionary psychology "severely limits the scope for any programme of reform," while historically, Singer "hideously caricature[s]" the so-called old left by "asserting that Marx denied the notion of human nature" (5), when, in fact, "Marx did not seek to deny the concept of human nature, merely to limit its restrictive power" by emphasizing "the continuity and interdependence of man and nature in a dialectical interaction" (66–67). David Stack, *The First Darwinian Left: Socialism and Darwinism, 1859–1914* (Cheltenham, U.K.: New Clarion Press, 2003).
7. Marx, qtd. in ibid., 74. Animals, of course, have "culture," too, in the sense that they modify their natural environment to enhance their chances of survival. A more precise definition of human labor and culture should tie these concepts to *techne*; see my discussion of Bernard Stiegler in chapter 3.
8. Based on a careful reading of Marx's sixth thesis in particular, Norman Geras already concluded back in 1983 that the "evidence is clear and abundant. Marx did not reject the idea of a human nature." At the same time, however, Geras was forced to admit that some Marxists still had a difficult time accepting this fact:

"Although the theory [of Marxism] requires, and in its present version explicitly possesses, a concept of human nature, many Marxists try nevertheless to repress it." See Geras, *Marx and Human Nature: Refutation of a Legend* (London: NLB, 1983), 89, 53.

9. Engels, qtd. in Singer, *Darwinian Left,* 27.
10. Engels, qtd. in Stack, *First Darwinian Left,* 1.
11. Geras, *Marx and Human Nature,* 95. "Any genuine materialism," Geras further explains, "must insist rather that human beings, for all that is distinctive about them as a species, and for all of their traits, activities and relationships which can only be explained by specificities of society and history, are nevertheless, like all other species, material and natural beings; 'irredeemably' rooted in a given biological constitution" (97).
12. Ibid., 24.
13. Cf. Diana Coole and Samantha Frost, eds., *New Materialisms: Ontology, Agency, and Politics* (Durham, N.C.: Duke University Press, 2010); Jane Bennett, *Vibrant Matter: A Political Ecology of Things* (Durham, N.C.: Duke University Press, 2010); Stacy Alaimo and Susan Hekman, "Introduction: Emerging Models of Materiality in Feminist Theory," in *Material Feminisms,* ed. Alaimo and Hekman (Bloomington: Indiana University Press, 2008), 1–19.
14. "Indeed, my own position is that theories of evolution and natural selection in biology had a social component before there was any question of reapplying them to social and political theory." Raymond Williams, "Social Darwinism," in *The Limits of Human Nature: Essays Based on a Course of Lectures Given at the Institute of Contemporary Arts, London,* ed. Jonathan Benthall (New York: Dutton, 1973), 86.
15. Lewontin, *Biology as Ideology,* 10. Inheritance, for example, was first and foremost a juridical concept before it became biologized. Prior to Darwin, the nineteenth-century use of the English word "*hereditary* refers primarily to the inheritance of property or title," whereas the noun *heredity* was not yet widely used at all. It was Darwin who seized on the latter in order to make the living organism itself "the vehicle of inheritance" such that "heredity is both nominalized and interiorized." See Keller, *Mirage of a Space between Nature and Nurture,* 21.
16. Williams, "Social Darwinism," 82.
17. Charles Darwin, *On the Origin of Species by Means of Natural Selection,* 6th ed. (New York: D. Appleton, 1982), 50. As Michael Ruse points out, Darwin "never thought that there must always be a literal struggle. He realized that this was as much a metaphor as anything else." See Ruse, "Darwinian Evolutionary Theory: Its Structure and Its Mechanism," in *The Oxford Handbook of Philosophy of Biology,* ed. Ruse, 36.
18. Charles Darwin, *The Descent of Man, and Selection in Relation to Sex,* 2nd ed. (New York: A. L. Burt, 1974), 152.
19. See Charles Darwin, *The Autobiography of Charles Darwin* (New York: Harcourt, Brace, 1958), 109, 395; qtd. in Mary Midgley, "Why Memes?" in *Alas, Poor Darwin: The Case against Evolutionary Psychology,* ed. Hilary Rose and Steven Rose (New York: Harmony, 2000), 77.

20. Monod, *Chance and Necessity,* xiii.
21. Ibid., 9.
22. Ibid., 178, 180.
23. Cf. ibid., 37, 150–51, 158.
24. Ibid., 7, 9.
25. Ibid., 9, 21.
26. Cf. Timothy Lenoir, *The Strategy of Life: Teleology and Mechanics in Nineteenth-Century German Biology* (Chicago: University of Chicago Press, 1982).
27. Ernst Mayr, *Toward a New Philosophy of Biology: Observations of an Evolutionist* (Cambridge, Mass.: Harvard University Press, 1988), 128.
28. Monod, *Chance and Necessity,* 24.
29. Ibid.
30. Cf. Thompson, *Mind in Life,* 167: "Perhaps the very first spontaneously self-assembling autopoietic systems were incapable of reproduction and therefore left no descendants." Compare to Monod, *Chance and Necessity,* xii: "So defined [as reproductive invariance], the theory of the genetic code constitutes the fundamental basis of biology."
31. Monod, *Chance and Necessity,* 13–15.
32. Ibid., 23, 24.
33. Ibid., 24.
34. Deacon, *Infinite Nature,* 307.
35. Cf. Monod, *Chance and Necessity,* 117, xii. Throughout his book, Monod, in fact, repeatedly links rather than separates invariance and teleonomy, which contradicts his central claim that "invariance necessarily precedes teleonomy." He claims, for example, that genetic mutations are acceptable to living systems only if they "do not lessen the coherence of the teleonomic apparatus." In passages like these, Monod argues that invariance is the *result* of teleonomic organization, not its *cause*: "It is the teleonomic apparatus, as it functions when a mutation first expresses itself, that lays down the essential initial condition for the admission . . . or rejection of the chance-bred innovative attempt" (119–20). On the "gene myth," see Hubbard and Wald, *Exploding the Gene Myth.*
36. Thompson, *Mind in Life,* 120.
37. Unlike his 1967 lecture, Monod's book from 1970 no longer mentions Teilhard de Chardin's concept of "noosphere" (i.e., the sphere of human thought), most likely because Althusser had focused his critique on the spiritualist connotations of this term.
38. Monod, *Chance and Necessity,* 126.
39. Ibid., 128.
40. Ibid., 129.
41. Ibid., 130; Wilson, *Consilience,* 150.
42. Ibid., 130, 162.
43. Cf. E. O. Wilson, *On Human Nature* (Cambridge, Mass.: Harvard University Press, 1978), 85.
44. Monod, *Chance and Necessity,* 127.

45. Ibid., 130, 162.
46. Wilson, *On Human Nature,* 80, 167.
47. Monod, *Chance and Necessity,* 163; Edward O. Wilson, *Sociobiology: The New Synthesis* (Cambridge, Mass.: Harvard University Press, 1975), 575, 562.
48. Ibid., 180.
49. Monod's inaugural lecture was expanded into a larger series of lectures (the Robbins Lectures) given in February 1969 at Pomona College in California. The Robbins Lectures provided the basis for Monod's subsequent book, *Le Hasard et la Necessité,* published in French in 1970 and translated into English a year later. My reading of Monod is based on this English version, though I will occasionally refer to both his inaugural lecture from 1967 and the original French version of the book from 1970.
50. Louis Althusser, "Philosophy and the Spontaneous Philosophy of the Scientists," in *Philosophy and the Spontaneous Philosophy of the Scientists and Other Essays,* ed. Gregory Elliott, trans. Ben Brewster et al. (London: Verso, 1990), 69–165. The last part (pp. 145–65) is titled "Appendix: On Jacques Monod" and translated by Warren Montag. Preceding the text is Althusser's assurance that the printed version reproduces the lecture "as delivered, without any alteration" (145).
51. Monod, *Chance and Necessity,* 40, 39.
52. Louis Althusser, *For Marx,* trans. Ben Brewster (New York: Vintage, 1969), 229.
53. Louis Althusser and Étienne Balibar, *Reading Capital,* trans. Ben Brewster (New York: Verso, 1979).
54. Louis Althusser, "The Humanist Controversy," in *The Humanist Controversy and Other Writings,* ed. François Matheron, trans. G. M. Goshgarian (London: Verso, 2003), 299n14.
55. Ibid., 273–74.
56. Ibid., 289; Althusser and Balibar, *Reading Capital,* 180.
57. Althusser, "Philosophy and the Spontaneous Philosophy of the Scientists," 133, 126.
58. Ibid., 142, 137, 139.
59. John Burdon Sanderson Haldane, *The Marxist Philosophy and the Sciences* (Freeport, N.Y.: Books for Libraries Press, 1969), 7.
60. Monod, *Chance and Necessity,* 39.
61. Ibid., 171.
62. Ibid., 43.
63. Gerhard Vollmer, *Biophilosophie* (Stuttgart: Reclam, 1995), 45; my translation.
64. Ibid., 152.
65. Monod, *Chance and Necessity,* 173.
66. Ibid., 165.
67. Althusser, "Philosophy and the Spontaneous Philosophy of the Scientists," 155.
68. Ibid., 145–46.
69. Althusser and Balibar, *Reading Capital,* 32, 317.
70. Althusser, "Philosophy and the Spontaneous Philosophy of the Scientists," 146.
71. Ibid., 165, 159. It should be noted that Monod himself readily accepts, and even

insists on, this idealist element in scientific research: "In science there is and will remain a Platonic element which could not be taken away without ruining it." See Monod, *Chance and Necessity,* 101.

72. Ibid., 151.
73. Louis Althusser, "Ideology and Ideological State Apparatuses," in *Norton Anthology of Theory and Criticism,* ed. Leitch et al., 1352.
74. For more details, see Thompson, *Mind in Life,* 417–41; and Deacon, *Incomplete Nature,* 143–81.
75. Althusser, "Philosophy and the Spontaneous Philosophy of the Scientists," 160, 152, 149, 153–54, 150.
76. Ibid., 152.
77. Ibid., 155.
78. Ibid.
79. Monod, *Chance and Necessity,* 165.
80. Althusser, "Philosophy and the Spontaneous Philosophy of the Scientists," 150–51.
81. Raymond Williams, "Social Darwinism," in *Culture and Materialism: Essential Writings,* trans. Jim McGuigan (London: Verso, 2005), 101.
82. Raymond Williams, "Problems of Materialism," in *Culture and Materialism,* ed. McGuigan, 112.
83. Althusser, "Philosophy and the Spontaneous Philosophy of the Scientists," 154.
84. Willard van Orman Quine, *Ontological Relativity and Other Essays* (New York: Columbia University Press, 1969), 48.
85. Ibid., 50, 48.
86. Richard Rorty, *Truth and Progress: Philosophical Papers* (Cambridge, U.K.: Cambridge University Press, 1998), 90.
87. "Realism is relative, determined by the system of representation standard for a given culture or person at a given time," Nelson Goodman writes. "Nothing is intrinsically a representation; status as representation is relative to symbol system. A picture in one system may be a description in another; and whether a denoting symbol is representational depends not upon whether it resembles what it denotes but upon its own relationships to other symbols in a given system." See Goodman, *Languages of Art,* 2nd ed. (Indianapolis: Hackett, 1976), 37, 226. Ian Hacking concurs: "We cannot reason as to whether alternative systems of reasoning are better or worse than ours, because the propositions to which we reason get their sense only from the method of reasoning employed. The propositions have no existence independent of the ways of reasoning towards them." See Hacking, *Historical Ontology,* 175. Hilary Putnam is a more ambivalent supporter of relativity, not least because he changed and continued to refine his views on realism throughout his career. A proponent of objectivism in his early years, Putnam reversed course around 1980 and advocated what he called "internal realism" instead. Simply put, internal realism states "that truth does not transcend use," meaning that different—even contradictory—statements referring to the same situation may each be true because the words are used differently and relate to different systems. Thus, although Putnam rejected Quine's notion of ontological

relativity, he did endorse Quine's view on norm-circularity. See Putman, *Representation and Reality* (Cambridge, Mass.: MIT Press, 1988), 115.

88. Bas C. van Fraassen, *The Scientific Image* (Oxford, U.K.: Clarendon Press, 1980), 18–19.
89. Andrew Pickering, *Constructing Quarks: A Sociological Study of Particle Physics* (Chicago: University of Chicago Press, 1984), 4–5. Similarly, see van Fraassen, *Scientific Image,* 156.
90. Pickering, *Constructing Quarks,* 405, 407.
91. Thomas S. Kuhn, *The Structure of Scientific Revolutions,* 3rd ed. (Chicago: University of Chicago Press, 1996), 150; my emphasis.
92. Cf. Godfrey-Smith, *Theory and Reality,* 177.
93. Cf. von Uexküll, *Foray into the World of Animals and Humans.*
94. Louis Althusser, *Philosophy of the Encounter: Later Writings, 1978–87,* ed. François Matheron and Oliver Corpet, trans. G. M. Goshgarian (London: Verso, 2006), 168, 169, 186.
95. Ibid., 254.
96. Althusser's reversals and self-contradictions are legendary and acknowledged by Althusser himself. See Louis Althusser, *Essays in Self-Criticism,* trans. Grahame Lock (New York: Humanities Press, 1976). G. M. Goshgarian's introduction to Althusser's posthumous publications provides a more detailed account. See Althusser, *Philosophy of the Encounter,* xiii–l. On the contradictory logic of Marxist philosophy in general, see Groys, *Communist Postscript.*
97. See Warren Montag, *Althusser and His Contemporaries: Philosophy's Perpetual War* (Durham, N.C.: Duke University Press, 2013), 173.
98. Unlike "rationalist materialism," Althusser insists, aleatory materialism is "not a Marxist philosophy, but a philosophy *for Marxism.*" See *Philosophy of the Encounter,* 259.
99. A more recent version of this belief in the radical contingency of the world is Quentin Meillassoux's *After Finitude: An Essay on the Necessity of Contingency,* trans. Ray Brassier (New York: Bloomsbury, 2010).

3. Cultural Evolution

1. Edward O. Wilson, *Sociobiology: The New Synthesis* (Cambridge, Mass.: Harvard University Press, 1975). Five years later, in an effort to make the book more affordable to the general public, Harvard published an abridged version comprising only 366 pages, about half its original size. See Edward O. Wilson, *Sociobiology: The Abridged Version* (Cambridge, Mass.: Harvard University Press, 1980). Citations and page numbers in my text refer to the original edition from 1975.
2. Given the intensity of the debate both inside and outside the academy, there is no shortage of analyses explaining how the scientific issues raised by *Sociobiology* intersected with larger political and moral issues discussed at the time. The most comprehensive account of the debate, including reprints from Darwin and other nineteenth-century writers, is provided in Arthur L. Caplan, ed., *The Sociobiology*

Debate: Readings on Ethical and Scientific Issues (New York: Harper, 1978). Two other excellent anthologies on the topic are Michael S. Gregory, Anita Silvers, and Diane Sutch, eds., *Sociobiology and Human Nature: An Interdisciplinary Critique and Defense* (San Francisco: Jossey-Bass, 1978); and George W. Barlow and James Silverberg, eds., *Sociobiology: Beyond Nature/Nurture? Reports, Definitions, and Debate* (Boulder, Colo.: Westview Press, 1980). The best monograph, in my view, is still Michael Ruse, *Sociobiology: Sense or Nonsense?* (New York: Springer, 1979).

3. The *New York Times*, for example, ran a front-page article titled "Sociobiology: Updating Darwin on Behavior" on March 28, 1975, followed by another major article just a few months later, on August 7, by C. H. Waddington, the great British philosopher and scientist whose work contributed to the rise of systems biology in the 1970s and 1980s. Both articles focused on Wilson's book and were mostly positive. Wilson himself wrote a lengthy essay on sociobiology in the *New York Times Magazine* on October 12.
4. The attack on Wilson is described in detail in his later book *Naturalist* (Washington, D.C.: Island Press, 2006), 338ff. Wilson's account of the event agrees with the report issued by Jon Beckwith and Bob Lange in *Science for the People* 10, no. 2 (1978): 38–40.
5. Sociobiology Study Group, "Against Sociobiology," *New York Review of Books*, November 13, 1975, http://www.nybooks.com/articles/1975/11/13/against-sociobiology/. Wilson's rebuttal, "For Sociobiology," was published in the same venue a month later, on December 11, 1975. See http://www.nybooks.com/articles/1975/12/11/for-sociobiology/.
6. Richard C. Lewontin, "Sociobiology: A Caricature of Darwinism," *PSA: Proceedings of the Biennial Meeting of the Philosophy of Science Association* 1976, no. 2 (1976): 22–23, 29–30.
7. Wilson, "For Sociobiology."
8. Wilson, *Sociobiology*, 547, 562, 17, 244, 368, 22.
9. Wilson, *Naturalist*, 339. In his *New York Review of Books* article from December 1975, Wilson had already denounced the actions of the Sociobiological Study Group as a "self-righteous vigilantism" that intimidates and "diminishes the spirit of free inquiry and discussion crucial to the health of the intellectual community." See also Edward O. Wilson, "Academic Vigilantism and the Political Significance of Sociobiology," in *The Sociobiology Debate*, ed. Caplan, 291–303; originally published in *BioScience* 26, no. 3 (1976): 187–90. See also Edward O. Wilson, "Introduction: What Is Sociobiology?, in *Sociobiology and Human Nature*, ed. Gregory, Silvers and Sutch, 1–12; Edward O. Wilson, "Foreword," in *Sociobiology Debate*, ed. Caplan, xi–xiv.
10. Wilson, "Foreword," xiii.
11. Wilson, "Introduction," 2.
12. Edward O. Wilson, "Sociobiology at Century's End," in *Evolution, Literature, and Film: A Reader*, ed. Brian Boyd, Joseph Carroll, and Jonathan Gottschall (New York: Columbia University Press, 2010), 99.

13. In a marvelous essay on the polemic history of evolutionary theory, David L. Hull makes the same argument, namely, that scientific discoveries and theories are inevitably judged in the context of larger social, political, and ethical concerns that may, at first, appear unrelated to the actual science yet "have proven time and again to be very real" throughout modern history. See Hull, "Sociobiology: Scientific Bandwagon or Traveling Medicine Show?" in *Sociobiology and Human Nature,* ed. Gregory, Silvers and Sutch, 148.
14. There is no mentioning of Althusser or Monod in either the scholarly or mainstream literature on the sociobiology debate from 1975 to the present. The same is true of Richard Levins and Richard C. Lewontin's book *The Dialectical Biologist* (Cambridge, Mass.: Harvard University Press, 1985).
15. Similar to Monod, Wilson begins and ends his book with philosophical reflections on the significance of evolutionary theory for human life, and like Monod he, too, adopts an exceedingly abstract, "extraterrestrial" perspective of "man" in order to mitigate the negative affective response he expects from his readers vis-à-vis a scientific theory that offends the entrenched anthropocentrism of the human species and hence "might be hard to accept" (Wilson, *Sociobiology,* 575). Wilson also echoed Monod's firm belief in the objective knowledge of science as the best means to quell this existential anxiety.
16. Wilson, *Naturalist,* 339.
17. Alexander J. Morin, "Revelation and Heresy in Sociobiology: A Review Essay," *Science, Technology, and Human Values* 4, no. 27 (1979): 25.
18. Wilson, *Sociobiology,* 575.
19. Ibid. The other evolutionary explanations for altruistic behavior are group selection (the phenotype sacrifices itself for the sake of the genotype), parental manipulation (the parents steer one of its offspring toward taking care of its siblings in order to increase their chances of survival and reproduction), and reciprocal altruism (a phenotype behaves altruistically in the hope that others will return the favor to the benefit of all).
20. According to Stephen Jay Gould, alternatives to the adaptationist program include (1) genetic drift; (2) "mechanically forced correlations," meaning that the "form of the part [at issue] is a correlated consequence of selection directed elsewhere"; and (3) the decoupling of selection and adaptation, because there exists selection without adaptation and adaptation without selection. Among the latter, there are numerous forms: "'Adaptation'—the good fit of organisms to their environment—can occur at three hierarchical levels with different causes" such as (a) "phenotypic plasticity" leading to physiological adaptations (which "are not heritable, though the capacity to develop them presumably is"); (b) cultural adaptation and learning; and (c) the Darwinian mechanism of selection on genetic variation. Further alternatives are (4) adaptations and selection, but no selective basis for differences among adaptations; and (5) adaptation and selection, but the adaptation is a secondary utilization of parts present for reasons of architecture, development, or history (spandrels, exaptation, etc.). Cf. Steven Rose, ed., *The Richness of Life: The Essential Stephen Jay Gould* (New York: Norton, 2007), 433–37.

21. Wilson, *Sociobiology*, 575.
22. Wilson, "Introduction," 1.
23. According to Hilary Rose and Steven Rose, genetic determinism claims, first, that "our biology is our destiny, written in our genes by the shaping forces of human evolution through natural selection and random mutation," and second, "that biotechnology, in the form of genetic engineering, can manipulate our genes in such a way as to rescue us from the worst of our fates." This, indeed, sums up the key points of Wilson's final chapter in *Sociobiology*. See Rose and Rose, "Introduction," in *Alas, Poor Darwin*, ed. Rose and Rose (New York: Harmony Books, 2000), 5.
24. Wilson, "Foreword," xiii–xiv.
25. Edward O. Wilson, *On Human Nature* (Cambridge, Mass.: Harvard University Press, 1978), x, 2, 88.
26. For a good summary of the Sokal affair, see "Sokal Affair," *Wikipedia*, last modified March 14, 2017, https://en.wikipedia.org/wiki/Sokal_affair.
27. See Rose and Rose, *Alas, Poor Darwin*. Contributors to this collection of essays came from a variety of disciplines (evolutionary theory, microbiology, psychology, anthropology, philosophy, etc.) and criticized EP for its "religious impulses" (the philosopher Dorothy Nelkin), its "uncritical embrace of Dawkinsian genetics" (the molecular biologist Gabriel Dover), its simplistic understanding of Darwinian theory that fails to include nonadaptive evolutionary changes (the paleontologist Stephen Jay Gould), and so forth. The shared premise of all contributors, however, was that EP had blatantly overstated its case and produced little more than "just-so" stories devoid of scientific evidence and impossible to falsify.
28. According to Rose and Rose, EP provides "a theory for everything" that not only lacks a solid scientific basis, but also leads to ethically and politically questionable, even dangerous, implications: "The political agenda of EP is transparently part of a right-wing libertarian attack on collectivity, above all the welfare state" (Rose and Rose, "Introduction," 9). Lewontin, too, claimed that sociobiology—and E. O. Wilson in particular—"totally confounds the properties and limitations of individuals with the properties and limitations of the social institutions that they create. It is the ultimate political manifestation of the belief that individual autonomous units determine the properties of the collectivities in which they assemble." See Lewontin, *Biology as Ideology*, 120–21.
29. Rose and Rose, "Introduction," 5.
30. Wilson, *Consilience*, 127–28.
31. John Tooby and Leda Cosmides, "Does Beauty Build Adapted Minds? Toward an Evolutionary Theory of Aesthetics, Fiction and the Arts," *SubStance* 94, no. 95 (2001): 15; my emphasis.
32. John Tooby and Leda Cosmides, "The Psychological Foundations of Culture," in *Adapted Mind*, ed. Barkow, Cosmides, and Tooby, 91.
33. Ibid., 110ff. See also David J. Buller, *Adapting Minds: Evolutionary Psychology and the Persistent Quest for Human Nature* (Cambridge, Mass.: MIT Press, 2005), 64ff.
34. Tooby and Cosmides, "Psychological Foundations of Culture," 99–100.
35. Some typical and well-known examples of EP's research agenda include David

Buss's claim that, in the Pleistocene era, it was advantageous for males to prefer nubile females and for females to prefer high-status males in order to secure the well-being of their offspring; or Martin Daly and Margo Wilson's claim that it was advantageous for parents to care most about their biological children (as opposed to foster children or those whose paternity remains unclear) in order to ensure the reproduction of their own genes; or John Tooby and Leda Cosmides's claim that we inherited a "cheater detection module" from our ancestors meant to weed out community members most likely to cheat and not keep their commitment to children and the family. See David M. Buss, *Evolutionary Psychology: The New Science of the Mind* (Boston: Allyn & Bacon, 1999), 105–29; Martin Daly and Margo Wilson, "A Reply to Gilles: Stepchildren *Are* Disproportionally Abused, and Diverse Forms of Violence *Can* Share Causal Factors," *Human Nature* 2, no. 4 (1991): 419–26; John Tooby and Leda Cosmides, "Cognitive Adaptations for Social Exchange," in *Adapted Mind,* ed. Barkow, Cosmides, and Tooby, 180–81. I shall examine the enormous influence of these studies in emerging fields like literary Darwinism in my next chapter, on evolutionary aesthetics.

36. Tooby and Cosmides, "Psychological Foundations of Culture," 87, 88.
37. What Darwin accomplished, Mayr argues, is "the switch from essentialist thinking to population thinking." See Mayr, *Toward a New Philosophy of Biology,* 15. The philosopher André Ariew, by contrast, insists that "Darwin's theory of natural selection, unlike its modern, post-Mendelian, synthesized version created in the early twentieth century, employs no statistical terms." See Ariew, "Teleology," in *The Cambridge Companion to the Philosophy of Biology,* ed. David L. Hull and Michael Ruse (Cambridge, U.K.: Cambridge University Press, 2008), 168.
38. Wilson, *On Human Nature,* 65; my emphasis.
39. Charles J. Lumsden and E. O. Wilson, *Genes, Mind, and Culture: The Coevolutionary Process* (Cambridge, Mass.: Harvard University Press, 1981), 13.
40. Robert Boyd and Peter J. Richerson, *The Origin and Evolution of Cultures* (Oxford, U.K.: Oxford University Press, 2005), 9.
41. Ibid., my emphasis.
42. Wilson, *Consilience,* 150.
43. Wilson, *On Human Nature,* 78.
44. Stephen Jay Gould, too, insists that "human cultural change operates fundamentally in the Lamarckian mode," which, due to its faster speed, continues to "overwhelm Darwinian rates of change." Hence he concludes, "Since Lamarckian and Darwinian systems work so differently, cultural change will receive only limited (and metaphorical) illumination from Darwinism." See Gould, in *Richness of Life,* ed. Rose, 464.
45. Lumsden and Wilson, *Genes, Mind, and Culture,* 2.
46. Wilson, "For Sociobiology." In *Sociobiology,* Wilson defined autocatalysis as "any process the rate of which is increased by its own products" (579). For example, once humans started to walk upright and were free to use their hands, the cultural impact on human natural development accelerated exponentially: "With mental capacity and the tendency to use artifacts increasing through mutual reinforce-

ment, the entire materials-based culture expanded. Now the species moved onto the dual track of evolution: genetic evolution by natural selection enlarged the capacity for culture, and culture enhanced the genetic fitness of those who made maximum use of it." See Wilson, *On Human Nature,* 85. In a later book from 2013, Wilson dates the beginning of cultural autocatalytic evolution quite a bit later, to about "60,000 to 50,000 years ago," and particularly emphasizes the beginning of agriculture some ten thousand years ago. See Edward O. Wilson, *The Social Conquest of Earth* (New York: Liveright, 2013), 91.

47. Monod, *Chance and Necessity,* 162.
48. Boyd and Richerson, *Origin and Evolution of Cultures,* 10; my emphasis.
49. Robert Boyd and Peter J. Richerson, *Not by Genes Alone: How Culture Transformed Human Evolution* (Chicago: University of Chicago Press, 2006), 9; my emphasis.
50. "Adult lactose digestion is controlled by a single dominant gene, and careful statistical work indicates that a history of dairying is the best predictor of a high frequency of this gene." Ibid., 192. Another example is the niche construction of many animals, such as beavers, who build a dam and thereby change the environmental conditions for all aquatic life in the region: "In this way of thinking, the products of culture become part of the selective environment of genes, just as the products of genes become part of the selective environment of culture" (277). See also Wilson, *Social Conquest of Earth,* 194.
51. Jean-Pierre Changeux, *The Physiology of Truth: Neuroscience and Human Knowledge,* trans. M. B. DeBevoise (Cambridge, Mass.: Harvard University Press, 2009), 37–38.
52. Keller, *Mirage of a Space between Nature and Culture,* 68, 80.
53. Boyd and Richerson, *Origin and Evoluiton of Cultures,* 53.
54. Michael Tomasello, *The Cultural Origin of Human Cognition* (Cambridge, Mass.: Harvard University Press, 1999), 15.
55. Ibid., 12.
56. Based on empirical studies that compare the cognitive development of human and nonhuman infants, Tomasello has argued all along that nonhuman primates "do not show the kind of flexibility of behavior of general causal principles characteristic of human children, from a fairly young age, as they try to solve physical problems. Nonhuman primates understand many antecedent-consequent relations in the world, but they do not seem to understand causal forces as mediating these relations" (ibid., 22). Posthumanists, no doubt, will want to challenge this distinction. But Tomasello explicitly warns against "anthropomorphizing or romanticizing the cognitive abilities of other animal species" (206), a danger I examine more closely in the coda of this book.
57. Cf. Leonhard Schilbach et al., "Toward a Second-Person Neuroscience," *Behavioral and Brain Sciences* 36 (2013): 393–462.
58. Tomasello, *Cultural Origin of Human Cognition,* 204.
59. Buller, *Adapting Minds,* 14.
60. Ibid., 419.
61. Ibid., 450.

62. "My target throughout this book," Buller states, "is Evolutionary Psychology *the paradigm*, and my criticisms are neither intended nor assumed to apply to all the work conducted with the *field of inquiry* that is sometimes called 'evolutionary psychology.' . . . This book is *not* a critique of the very idea that human psychology can be understood from an evolutionary perspective" (*Adapting Minds*, 12).
63. Richard Dawkins, *The Selfish Gene* (1976; repr., Oxford, U.K.: Oxford University Press, 1996), 192.
64. Cf. ibid., 199–200.
65. Susan Blackmore, *The Selfish Meme* (Oxford, U.K.: Oxford University Press, 1999), 27, 31, 235. As evidence for this claim, Blackmore points to the development of agriculture: "we know from the skeleton of early farmers that they were malnourished and sickly" (27), meaning that it would have been better for our ancestors to remain hunter–gatherers—at least from the (selfish) gene's point of view. Other examples Blackmore cites are the development of language, our big brains, and human consciousness, all of which were maladaptive from the gene's point of view.
66. Kate Distin, *The Selfish Meme: A Critical Reassessment* (Cambridge, U.K.: Cambridge University Press, 2004), 16; Robert Aunger, *The Electric Meme: A New Theory of How We Think* (2002; repr., New York: Free Press, 2010), 208. To make his case, Aunger distinguishes between meme theories rooted in epidemiology (e.g., memes conceptualized as viruses) and meme theories rooted in replication (e.g., memes conceptualized as genes). He suggests that the former is more productive and sensible an approach than the latter: "If memes are parasites," Aunger concludes, "they must be symbiotic ones."
67. Charles J. Lumsden and E. O. Wilson, "The Relation between Biological and Cultural Evolution," *Journal of Social and Biological Structures* 8 (1985): 343–59, 350.
68. Lumsden and Wilson, *Genes, Mind, and Culture*, 372; see also 16–17.
69. Lumsden and Wilson, "Relation between Biological and Cultural Evolution," 343.
70. Kate Distin, for example, defines memes exclusively on the message level as some kind of representational *content*: "Such a theory must be able to determine both which sorts of representations count as memes, and how we can specify the content of those that do." See Distin, *Selfish Meme*, 20. According to Robert Aunger, by contrast, memes are neither abstract ideas nor representational contents of some sort. Instead, memes are chemical–electronic *signals* that use bodies and artifacts as hosts in order to replicate: "A meme, then, is essentially the state of a node in a neuronal network capable of generating a copy of itself in either the same or a different neuronal network, without being destroyed in the process." See Aunger, *Electric Meme*, 325–26. Both of these approaches are plausible and rich in detail, and both are worth pursuing, in my view.
71. Richard Dawkins, *The Extended Phenotype: The Long Reach of the Gene* (Oxford, U.K.: Oxford University Press, 1982), 109. See also Dennett, *Darwin's Dangerous Idea*, 344.
72. Schrödinger, *What Is Life?*, 22.
73. Evelyn Fox Keller emphasizes again and again that "by themselves, the entities we call genes do not act; they do not have agency." See Keller, *Mirage of a Space*

between Nature and Culture, 6. Instead, genes interact with their environment from the very beginning of organic development, and neither "part" functions at all without the other. This is why "the very notion of a gene as an autonomous element, as an entity that exists in its own right, is a fiction." Similarly, see Hubbard and Wald, *Exploding the Gene Myth*; and Lewontin, *It Ain't Necessarily So.*

74. In his otherwise excellent book, the science journalist Ed Regis still advocates the central dogma of DNA: "Because it is the core entity of life, controlling heredity, growth, and practically every other biologically significant function or factor, DNA's significance cannot be exaggerated. DNA is to life what atoms are to matter." It is precisely this misleading analogy between physics and the life sciences that I want to challenge in this book. See Regis, *What Is Life?,* 51.
75. As Judith Roof argues, DNA discourse is a symptom of the reductionist logic that has dominated the West since modern times: "DNA is the promise of a long legacy, the coming together of two related tendencies—the reductionist analytic atomism of 'hard' science and the structuralism of philosophy, narrative, and cultural criticism." These deep philosophical roots of DNA discourse, I believe, have fortified its resilience against scholarly critique, the kind of resilience meme discourse was unable to match. See Roof, *Poetics of DNA,* 33.
76. Richard Dawkins, *The Blind Watchmaker: Why the Evidence of Evolution Reveals a Universe without Design* (New York: Norton, 1986), 196.
77. Dawkins, *Extended Phenotype,* 109.
78. Dennett, *Darwin's Dangerous Idea,* 344.
79. Richard Dawkins, "Foreword," in Susan Blackmore, *The Meme Machine* (Oxford, U.K.: Oxford University Press, 1999), xiv.
80. Ibid., ix.
81. David Deutsch, *The Fabric of Reality: The Science of Parallel Universes and Its Implications* (London: Penguin, 1997), 406.
82. Blackmore, *Selfish Meme,* 17; my emphasis.
83. Boyd and Richerson, *Origin and Evolution of Cultures,* 421.
84. Boyd and Richerson, *Not by Genes Alone,* 61–62.
85. Deutsch, *Fabric of Reality,* 41.
86. Boyd and Richerson, *Origin and Evolution of Cultures,* 334.
87. Ibid., 311, 326.
88. Lumsden and Wilson, "Relation between Biological and Cultural Evolution," 347.
89. Ibid.
90. Ibid.
91. Wilson, *Consilience,* 134.
92. Cf. Lumsden and Wilson, *Genes, Mind, and Culture,* 7.
93. Lumsden and Wilson, "Relation between Biological and Cultural Evolution," 349.
94. Charles Lumsden and E. O. Wilson, *Promethean Fire* (Cambridge, Mass.: Harvard University Press, 1983), 121. See also Lumsden and Wilson, *Genes, Mind, and Culture,* 368.
95. Lumsden and Wilson, *Genes, Mind, and Culture,* 27, 29.
96. Ibid., 28–29.

97. Lumsden and Wilson, "Relation between Biological and Cultural Evolution," 348.
98. Ibid.
99. Lumsden and Wilson, *Genes, Mind, and Culture,* 27.
100. Ibid., 368.
101. Midgley, "Why Memes?," 80. Rose and Rose, too, argue that neither thought nor culture "are granular or can be decomposed into memic units. Unlike genes, memes are content-free concepts." See Rose and Rose, "Introduction," 12.
102. Cf. Terrence Deacon, *Incomplete Nature: How Mind Emerged from Matter* (New York: Norton, 2013).
103. Boyd and Richerson, *Origin and Evolution of Cultures,* 327.
104. Boyd and Richerson, *Not by Genes Alone,* 73.
105. Ibid., 79.
106. Boyd and Richerson, *Origin and Evolution of Cultures,* 326.
107. Ibid., 421.
108. Bernard Stiegler, *Technics and Time 1: The Fault of Epimetheus,* trans. Richard Beardsworth and George Collins (Stanford, Calif.: Stanford University Press, 1998), 174. See also Bernard Stiegler, *Technics and Time 2: Disorientation,* trans. Stephen Barker (Stanford, Calif.: Stanford University Press, 2009); and Bernard Stiegler, *Technics and Time 3: Cinematic Time and the Question of Malaise,* trans. Stephen Barker (Stanford, Calif.: Stanford University Press, 2011). Stiegler's overall theory is interdisciplinary, comprehensive, and complex; it includes references to anthropological and scientific studies as well as a detailed account of Western philosophical ideas about human nature (from Aristotle to Derrida via Hume, Rousseau, Kant, Husserl, and Heidegger) explicated across all three volumes of *Technics and Time,* with a fourth volume scheduled to appear in the near future. Obviously, I cannot provide a detailed account of Stiegler's various arguments at this point; I can only address those I consider most relevant for my discussion of cultural evolution as developed in this chapter so far. That said, I do think it might be helpful for readers to get a sense of the overall trajectory of Stiegler's theory as summarized—all too coarsely, I am afraid—in the following set of theses: (1) The quest for human nature and its historical origin is, literally, based on a fiction. It is a philosophical narrative exemplified in Rousseau and rooted in our cognitive inability to conceive of the essence of humanity in other than self-contradictory terms. (2) The human is an essentially prosthetic being defined by the development and use of exterior memory devices. The being of humans is inextricably intertwined and coevolving with technics. (3) The human brain is shaped by the constant process of uploading and downloading memory from and to external memory devices (i.e., technics). The brain evolves by reinteriorizing the cognitive power gained though the exteriorization of memory. (4) Technics is self-evolving and leads to the artificially accelerated evolution of human culture that is both analogous to, yet separate from, biological evolution. (5) The nineteenth-century invention of modern mnemotechnical apparatus (phonograph, photography, cinematography, etc.) gives rise to a new kind of memory besides genetic and somatic memory—namely, what Stiegler calls "epiphylogenetic memory."

(6) The structure of modern human consciousness is thoroughly cinematographic. (7) Today's globalized capitalism represents a new phase of technogenesis that threatens to reorganize/colonize the human mind under the auspices of profit maximization.

109. Stiegler defines technics—as distinct from technology—with reference to the Aristotelian term *tekhne,* usually translated as art, craft, or rhetorical skill (i.e., sophistry). Yet Stiegler also insists that technics is a far broader and more abstract term than *tekhne* because it designates the entire milieu of human culture as a distinct niche in the human environment. "Far from being reducible to the physical, technics is thus a milieu that conditions the temporality of 'practical reason,' of the will." See Stiegler, *Technics and Time 3,* 196.
110. Stiegler, *Technics and Time 1,* 177.
111. Stiegler, *Technics and Time 2,* 4.
112. Ibid., 99.
113. Siegfried Zielinski, *Deep Time of the Media: Toward an Archaeology of Seeing and Hearing by Technical Means,* trans. Gloria Custance (Cambridge, Mass.: MIT Press, 2008).
114. Stiegler, *Technics and Time 1,* 27, 92, 35; similarly, see 44–45.
115. Ibid., 176.
116. Ibid., 149.
117. Stiegler, *Technics and Time 2,* 99.
118. Varela himself was strongly opposed to this move, which he considered an illegitimate metaphorical extension of autopoiesis from living organisms to social structures. See Protevi, *Life, War, Earth,* 25ff.
119. Stiegler, *Technics and Time 1,* 175.
120. Ibid., 137.
121. Vicki Kirby, *Quantum Anthropologies: Life at Large* (Durham, N.C.: Duke University Press, 2011), 83.
122. "What we do in the humanities and social sciences," Kirby claims, "is not an idle metaphor," because "so-called cultural critics are already practicing science—we really are in the business of material production" (xi).
123. Ibid., ix–x.
124. Jerome A. Feldman, *From Molecule to Metaphor: A Neural Theory of Language* (Cambridge, Mass.: MIT Press, 2006), 199.
125. George Lakoff and Mark Johnson, *Philosophy in the Flesh: The Embodied Mind and Its Challenge to Western Thought* (New York: Basic Books, 1999), 59. See also George Lakoff and Mark Johnson, *Metaphors We Live By,* 2nd ed. (Chicago: University of Chicago Press, 2003). In both books, Lakoff and Johnson provide numerous examples of what they call the "primary metaphors" that organize our conceptual toolkit, such as the metaphor "More Is Up": "In More Is Up, a subjective judgment of quantity is conceptualized in terms of the sensorimotor experience of verticality." Human cognition thus establishes a "correspondence between quantity and verticality" that "arises from a correlation in our normal everyday experiences, like pouring more water into the glass and seeing the level go up." See Lakoff and

Johnson, *Philosophy in the Flesh,* 47. Other examples of primary metaphors are "Knowing Is Seeing," "Important Is Big," and "Similarity Is Closeness" (50).

126. Cf. Mark Johnson, *The Meaning of the Body: Aesthetics of Human Understanding* (Chicago: University of Chicago Press, 2007). Johnson develops a pragmatist notion of meaning as "a matter of relations and connections grounded in bodily organism-environment coupling, or interaction. The meaning of something is its relations, actual and potential, to other qualities, things, events, and experiences" (265). Although Johnson does not mention Derrida, his definition of "the meaning of meaning" seems more than compatible with the differential logic of deconstruction. It also works well with the sense of bioaesthetics outlined in this book.
127. For a concise history of the contested English translation of this phrase, see the editors' introduction to Derrida in Leitch et al., ed., *Norton Anthology of Theory and Criticism,* 1816.

4. Evolutionary Aesthetics

1. Weitz, qtd. in *Ästhetische Grundbegriffe,* ed. Barck et al., 1:576; my translation.
2. Schaeffer, *Art in the Modern Age,* 6.
3. Boris Groys, *Introduction to Anti-Philosophy* (New York: Verso, 2012), 73. Similarly, see Peter Bürger, *Theory of the Avant-Garde,* trans. Michael Shaw (Minneapolis: University of Minnesota Press, 1984).
4. Cf. George Dickie, *Art and the Aesthetic: An Institutional Approach* (Ithaca, N.Y.: Cornell University Press, 1974).
5. Cf. Hans Belting and Andrea Buddensieg, *The Global Art World: Audiences, Markets, and Museums* (Ostfildern: Hatje Cantz, 2009).
6. E. O. Wilson, "On Art," in *Biopoetics,* ed. Cooke and Turner, 77.
7. Steven Pinker, *The Blank Slate: The Modern Denial of Human Nature* (London: Penguin, 2003), 404.
8. Dissanayake, *Homo Aestheticus,* xix.
9. The proponents of evolutionary aesthetics are rather vague on this point as well. Dennis Dutton dates the environment of evolutionary adaptation (EEA) to about one hundred thousand years ago. See Dutton, *The Art Instinct: Beauty, Pleasure, and Human Evolution* (New York: Bloomsbury, 2009) 8. On the other hand, Dissanayake speaks of "several hundred thousand years." See Dissanayake, *Homo Aestheticus,* xii. This vague time frame for EEA has emerged as a major point of criticism against evolutionary psychology in general, succinctly summarized by McKinnon: "Among the many problems with their conjectures about evolutionary origins, two stand out: a remarkable vagueness about the time, place, and circumstances of the environment of evolutionary adaptation; and an equally remarkable specificity about the nature of social relations in that hypothetical time and place." Susan McKinnon, *Neo-Liberal Genetics: The Myths and Moral Tales about Evolutionary Psychology* (Chicago: Prickly Paradigm Press, 2006), 136–37. Similarly, see Jonathan Kramnick, "Against Literary Darwinism," *Critical Inquiry* 37, no. 2 (2011): 315–47.
10. Dutton, *Art Instinct,* 14.

11. A similar argument was already made by Gordon H. Orians and Judith H. Heerwagen in "Evolved Responses to Landscapes," in *Adapted Mind,* ed. Barkow, Cosmides, and Tooby, 555–78.
12. Dutton, *Art Instinct,* 62, 76.
13. Cf. V. S. Ramachandran and William Hirstein, "The Science of Art: A Neurological Theory of Aesthetic Experience," *Journal of Consciousness Studies* 6, nos. 6–7 (1999): 15–51.
14. John Dewey, *Art as Experience* (London: Penguin, 2005), 95.
15. Theodor W. Adorno, *Aesthetic Theory,* trans. Robert Hullot-Kentor (Minneapolis: University of Minnesota Press, 1988), 114.
16. Dissanayake, *Homo Aestheticus,* 42, 52.
17. Cf. Kathryn Coe, *The Ancestress Hypothesis: Visual Art as Adaptation* (New Brunswick, N.J.: Rutgers University Press, 2003). See also the excellent, though still not translated study by Karl Eibl, *Animal Poeta: Bausteine der biologischen Kultur- und Literaturtheorie* (Paderborn: Mentis, 2004).
18. Ellen Dissanayake, "The Arts after Darwin: Does Art Have an Origin and Adaptive Function?," in *World Art Studies: Exploring Concepts and Approaches,* ed. K. Zijlmans and W. van Damme (Amsterdam: Valiz, 2008), 244.
19. Gilgen, "Aesthetics," 22.
20. See Anjan Chatterjee, *The Aesthetic Brain: How We Evolved to Enjoy Beauty and Enjoy Art* (Oxford, U.K.: Oxford University Press, 2015).
21. Ibid., 142.
22. Dutton, *Art Instinct,* 63.
23. William K. Wimsatt Jr. and Monroe C. Beardsley, "The Intentional Fallacy," in *Norton Anthology of Theory and Criticism,* ed. Leitch et al., 1235.
24. Cf. Wilhelm Dilthey, *Selected Works, Volume I: Introduction to the Human Sciences,* ed. Rudolf A. Makkreel and Frithjof Rodi (Princeton, N.J.: Princeton University Press, 1991); Franco Moretti, *Graphs, Maps, Trees: Abstract Models for a Literary History* (London: Verso, 2005).
25. Northrop Frye, "The Archetypes of Literature," in *Norton Anthology of Theory and Criticism,* ed. Leitch et al., 1306.
26. Ibid., 1309.
27. See the editors' comments on Frye in Leitch et al., eds., *Norton Anthology of Literary Criticism,* 1301–3.
28. Carroll, *Reading Human Nature,* 77, 84.
29. Brett Cooke, "Edward O. Wilson on Art," in *Biopoetics,* ed. Cooke and Turner, 109.
30. Brian Boyd, Joseph Carroll, and Jonathan Gottschall, "Introduction," in *Evolution, Literature, and Film,* ed. Boyd, Carroll, and Gottschall, 13.
31. Joseph Carroll, Jonathan Gottschall, John A. Johnson, and Daniel J. Kruger, *Graphing Jane Austen: The Evolutionary Basis of Literary Meaning* (London: Palgrave, 2012), 160.
32. Carroll's own view is clear enough: "Foucault champions an irrationalism that has deep affinities with doctrinaire authoritarianism." See Carroll, "Rejoinder to the Responses," *Style* 42, nos. 2–3 (2008): 310. This outlandish statement is all the more

baffling given Foucault's argumentative rigor and the fact that he never advocated irrationalism as a guiding principle for (his) philosophy. What he *did* do, admittedly, was to probe the epistemological limits of modern rationalism in order to expose the normative assumptions that inform the production of scientific knowledge. To dismiss the epistemological questions Foucault raises as irrational, doctrinaire, or authoritarian is simply being obtuse.

33. Joseph Carroll, Jonathan Gottschall, John Johnson, and Daniel Kruger, "Imagining Human Nature," in *Evolution, Literature, and Film*, ed. Boyd, Carroll, and Gottschall, 214.
34. "To imagine that Foucault rediscovered, came back to the subjectivity he'd initially rejected, is as fundamental a misunderstanding as the one about 'the death of man.' Indeed, I think subjectivification has little to do with any subject. It's to do, rather, with an electric or magnetic field, an individuation taking place through intensities (weak as well as strong ones), it's to do with individuated fields, not persons or identities." Gilles Deleuze, *Negotiations, 1972–1990*, trans. Martin Joughin (New York: Columbia University Press, 1996), 93.
35. Cf. Coole and Frost, *New Materialisms*; Cary Wolfe, *What Is Posthumanism?* (Minneapolis: University of Minnesota Press, 2010).
36. Boyd, Carroll, and Gottschall, "Introduction," 12.
37. Wilson, "Evolutionary Social Constructivism," 29.
38. Carroll et al., "Imagining Human Nature," 215.
39. Tony Davies, *Humanism* (London: Routledge 1997), 141.
40. Wilson, "Foreword from the Scientific Side," viii. One of the most emphatic rejections of this claim and the neo-Darwinian view of human nature is Marshall Sahlins, *The Western Illusion of Human Nature* (Chicago: Prickly Paradigm Press, 2008).
41. Patrick Colm Hogan, for example, has made a compelling case for the nondiscriminatory nature of universal cultural categories: "A universalist program that succeeds in uncovering genuinely universal principles of human thought and human society, principles that are not relative to race or culture, necessarily runs contrary to racism and ethnocentrism." That sounds right, except that everything depends on the specific use made of such universalist categories in particular historical contexts, as I argue throughout this book. Hogan, "Literary Universals," in *Introduction to Cognitive Literary Studies*, ed. Lisa Zunshine (Baltimore: Johns Hopkins University Press, 2010), 60.
42. See Jonathan Gottschall, *The Rape of Troy: Evolution, Violence and the World of Homer* (Cambridge, U.K.: Cambridge University Press, 2008), 89.
43. Joseph Carroll, "An Evolutionary Paradigm for Literary Studies," *Style* 42, nos. 2–3 (2008): 103–34. See also *Literary Darwinism* and *Reading Human Nature*.
44. Ellen Spolsky, for example, strongly rejects Carroll's version of literary Darwinism and its notion of human nature. "It is, in fact, ludicrous to think that in the project of describing human nature, the novels of Jane Austen are needed to confirm the understanding that mating is crucial for the continuation of the species!" See Spolsky, "The Centrality of the Exceptional in Literary Study," *Style* 42, nos. 2–3 (2008):

288. Similarly, Eugene Goodheart claims that Carroll's "chutzpah is breathtaking, given the absence of anything approaching a theory of literature in the paradigm he presents." See Goodheart, "Do We Need Literary Darwinism?," *Style* 42, nos. 2–3 (2008): 182. Roger Seaman sums up the complaints of many literary scholars vis-à-vis Carroll's evolutionary paradigm for the study of culture: "The program of Darwinian literary studies . . . encompasses a number of projects whose relationships are unclear, whose ambitions are unrealizable, and whose interpretative power is weak. However, the most disturbing aspect of the project is that it depends on misunderstanding the nature of literary studies." See Seaman, "Literary Darwinism as Science and Myth," *Style* 42, nos. 2–3 (2008): 261.

45. Carroll, "Evolutionary Paradigm for Literary Studies," 107.
46. Cf. Geoffrey Miller, *The Mating Mind: How Sexual Choice Shaped the Evolution of Human Nature* (Norwell, Mass.: Anchor, 2001). Feminist scholars, too, have recently emphasized the distinction between natural and sexual selection in their interpretation of Darwinian theory. See Grosz, *Becoming Undone.*
47. For a detailed critique of this view, see Boyd and Richerson, *Not By Genes Alone,* 74–76; Carroll, *Literary Darwinism,* xxi; and Carroll, *Reading Human Nature,* 66.
48. Dissanayake, for example, argues that "humans universally need art" because the "behavior of art had 'selective' or 'survival' value: it was a biological necessity." See Dissanayake, *Homo Aestheticus,* 34–35. She readily concedes, however, that art no longer serves this purpose today. Dutton agrees: the "ensemble of adaptations that become the human art instinct go back in our prehistory only a hundred thousand years or so," but are no longer adaptive in today's environment. See Dutton, *The Art Instinct,* 8. Brian Boyd likewise argues that "art, too, is a specifically human adaptation, biologically part of our species." See Boyd, *On the Origin of Stories: Evolution, Cognition, and Fiction* (Cambridge, Mass.: Harvard University Press, 2010), 1. Jonathan Gottschall, finally, remains agnostic on this issue: "I don't know for sure whether storytelling is an evolutionary adaptation or a side effect, and neither at this point does anyone else." See Gottschall, *The Storytelling Animal: How Stories Make Us Human* (Boston: Harcourt, 2012), 30. I share Gottschall's agnosticism; it is clearly the most reasonable stance to take on the matter.
49. Steven Pinker, *How the Mind Works* (New York: Norton, 2009), 534, 526.
50. Faced with a severe critique of his position by Carroll and others, Pinker responded "that virtually all of these critics misunderstand the meaning of adaptation in the biologist's sense. The term does not refer to a trait that is healthy, valuable, edifying, or uplifting but rather to a trait that increased the number of surviving offspring in an organism's ancestral lineage. As such, the critics are really trying to affirm the moral or spiritual or emotional or scientific importance of music, rather than applying the scientific criteria for adaptation." Ibid., x.
51. Cf. Joseph Carroll, "An Open Letter to Jonathan Kramnick," *Critical Inquiry* 38, no. 2 (2012): 405–10. Similarly, see Brian Boyd, "For Evocriticism: Minds Shaped to Be Reshaped," *Critical Inquiry* 38, no. 2 (2012): 394–404.
52. "Darwin's dangerous idea is reductionism incarnate, promising to unite and explain just about everything in one magnificent vision. Its being the idea of an

algorithmic process makes it all the more powerful, since the substrate neutrality it thereby possesses permits us to consider its application to just about anything." See Dennett, *Darwin's Dangerous Idea,* 82.

53. Wilson, *Consilience,* 266.
54. Carroll, *Literary Darwinism,* 84, 72, vii.
55. Carroll, "Evolutionary Paradigm for Literary Studies," 105. If anything, Carroll's research program has become even more totalizing since then. Cf. Carroll, *Reading Human Nature,* 70.
56. Jacques Monod repeatedly emphasized the statistical nature of post-Newtonian, scientific knowledge. Today's theories "anticipate the existence, the properties, the interrelations of certain *classes* of objects or events," yet they are "obviously not . . . able to foresee the existence or the distinctive characteristics of any *particular* object or event" (43). At stake, for Monod, is the "very real sense [in which] the organism does effectively transcend physicals—even while obeying them" (80). So, at no point did Monod proclaim the nullity of scientific laws; on the contrary, the unpredictability of a particular organism's evolutionary development actually confirms these laws. See Monod, *Chance and Necessity.*
57. Carroll, *Reading Human Nature,* 84.
58. Cf. Carroll, "Evolutionary Paradigm for Literary Studies," 105.
59. Carroll, "Rejoinder to the Responses," 324. Steven Pinker, too, rejects what he calls the doctrine of "holistic interactionism": "The everything-affects-everything diagram turns out to be not sophisticated but dogmatic." See Pinker, "Why Nature and Nurture Won't Go Away," *Daedalus* 133, no. 4 (2004): 17.
60. Carroll, "Rejoinder to the Responses," 325.
61. Cf. Wilson, "Evolutionary Social Constructivism," 20–37. Ellen Spolsky, too, has pointed out "that Darwin's theory of evolution is significantly homologous to the post-structuralist critique of representation." See Spolsky, "Darwin and Derrida: Cognitive Literary Theory as a Species of Post-Structuralism," *Poetics Today* 23, no. 1 (2002): 43.
62. Carroll, "Rejoinder to the Responses," 324.
63. Carroll, *Reading Human Nature,* 44.
64. Ibid., 44–45.
65. Friedrich A. von Hayek, *New Studies in Philosophy, Politics, Economics and the History of Ideas* (London: Routledge, 1976), 291.
66. Geoffrey M. Hodgson, "Is Social Evolution Lamarckian or Darwinian?" in *Darwinism and Evolutionary Economics,* ed. John Laurent and John Nightingale (Cheltenham, U.K.: Edward Elgar, 2001), 100. "It is quite wrong, therefore, for evolutionary economists to distance themselves completely from either Darwinism or Lamarckism," Hodgson continues. "In general, and broadly interpreted, the two are compatible" (116).
67. "Darwinian principles [of evolution] are more fundamental, because Lamarckism itself always relies on Darwinian props." See ibid., 116. In my view, however, this fact does not amount to the significantly stronger claim that "Darwinian evolution is necessary to *guide* Lamarckian evolution." See Thorbjørn Knudsen, "Nest-

ing Lamarckism within Darwinian Explanations," in *Darwinism and Evolutionary Economics,* ed. Laurent and Nightingale, 129; my emphasis.

68. Hodgson, "Lamarckian or Darwinian?," 105.
69. Ibid., 102.
70. Michael Ruse, *Taking Darwin Seriously: A Naturalistic Approach to Philosophy* (New York: Prometheus, 1998), 35, 54.
71. Vollmer, *Biophilosophie,* 148; my translation.
72. Boyd, Carroll, and Gottschall, "Introduction," in *Evolution, Literature, and Film,* ed. Boyd, Carroll, and Gottschall, 1.
73. Kramnick, "Against Literary Darwinism," 345. See also Jonathan Kramnick, "Literary Studies and Science: A Response to My Critics," *Critical Inquiry* 38, no. 2 (2012): 430–60. Another excellent, yet less influential critique of literary Darwinism can be found in Frank Kelleter, "A Tale of Two Natures: Worried Reflections on the Study of Literature and Culture in an Age of Neuroscience and Neo-Darwinism," *Journal of Literary Theory* 1, no. 1 (2007): 153–89.
74. Kramnick, "Against Literary Darwinism," 345.
75. Cf. Boyd, "For Evocriticism"; Jonathan Gottschall, "Toward Consilience, Not Literary Darwinism," *Scientific Study of Literature* 3, no. 1 (2013): 16–18.
76. Carroll, *Literary Darwinism,* 133.
77. Cf. Geoffrey Harpham, *Language Alone: The Critical Fetish of Modernity* (New York: Routledge, 2002).
78. Again, Brian Boyd's work is exemplary in that regard. Besides *The Origin of Stories,* see also Boyd's essay "Art and Evolution: The Avant-Garde as Test Case," in *Evolution, Literature, and Film,* ed. Boyd, Carroll, and Gottschall, 433–54.
79. Joseph Carroll, *Evolution and Literary Theory* (Columbia: University of Missouri Press, 1994), 145, 136, 137.
80. Ibid., 68.
81. Carroll, *Reading Human Nature,* 108.
82. Carroll et al., "Imagining Human Nature," 218.
83. Gottschall, *Storytelling Animal,* 4, 3.
84. Friedrich Kittler, *Discourse Networks, 1800/1900,* trans. Michael Metteer (Stanford, Calif.: Stanford University Press, 1992).
85. Kurt Schwitters, "An Anna Blume," in *Deutsche Gedichte: Von den Anfängen bis zur Gegenwart,* ed. Elisabeth K. Paefgen and Peter Geist (Berlin: Cornelsen, 2010), 623–24; my translation.
86. The idiomatic meaning of the German original *Anna Blume hat einen Vogel* is "Anna Blume is cuckoo."
87. Roland Barthes, "The Death of the Author," in *Norton Anthology of Literary Criticism,* ed. Leitch et al., 1323, 1325, 1326.
88. Ibid., 1325.
89. Carroll, *Literary Darwinism,* 133.
90. Nancy E. Aiken, *The Biological Origins of Art* (Westport, Conn.: Praeger, 1998), 7.
91. As Brett Cooke emphasizes, biopoetics "use[s] the newly available tools of sociobiology to dramatically increase our understanding of the centrality and persistence

of the arts in human experience." Brett Cooke, "Biopoetics: The New Synthesis," in *Biopoetics,* ed. Cooke and Turner, 3.

92. For a brief yet by no means exhaustive overview of pertinent examples, see Kelleter, "Tale of Two Natures," 160–61; and Kramnick, "Literary Studies and Science," 443–44.
93. David P. Barash and Nanelle R. Barash, *Madame Bovary's Ovaries: A Darwinian Look at Literature* (New York: Delacorte Press, 2005), 101.
94. Cf. Gertrude Stein, "Sacred Emily," *Genius,* n.d., https://genius.com/Gertrude-stein-sacred-emily-annotated.
95. "All organisms that have ever lived—" Richard Dawkins astutely observes, "every animal and plant, all bacteria and all fungi, every creeping thing, and all readers of this book—can look back at their ancestors and make the following proud claim: Not a single one of our ancestors died in infancy." Richard Dawkins, "The Digital River," in *Evolution, Literature, and Film,* ed. Boyd, Carroll, and Gottschall, 55. I certainly hope that Dawkins's readers ended up with more and better insights than that.
96. Joseph Carroll, Jonathan Gottschall, John A. Johnson, and Daniel Kruger, "Paleolithic Politics in British Novels of the Longer Nineteenth Century," in *Evolution, Literature, and Film,* ed. Boyd, Carroll, and Gottschall, 504.
97. Ibid.
98. Joseph Carroll, "The Cuckoo's History: Human Nature in *Wuthering Heights,*" in *Evolution, Literature, and Film,* ed. Boyd, Carroll, and Gottschall, 317.
99. Carroll et al., "Paleolithic Politics in British Novels of the Longer Nineteenth Century," 506.
100. Daniel J. Kruger, Maryanne Fisher, and Ian Jobling, "Proper Hero Dads and Dark Hero Cads: Alternate Mating Strategies Exemplified in British Romantic Literature," in *Literary Animal,* ed. Gottschall and Wilson, 237.
101. Paul Bloom, "Who Cares about the Evolution of Stories?," *Critical Inquiry* 38, no. 2 (2012): 388–94.
102. Jonathan Gottschall, "Quantitative Literary Study: A Modest Manifesto and Testing the Hypothesis of Feminist Fairly Tale Studies," in *Literary Animal,* ed. Gottschall and Wilson, 219.
103. Katja Mellmann rightly points out that Gottschall "claims to be testing a literary–critical hypothesis, but in fact he is testing an anthropological hypothesis using literary data." This methodological confusion, she claims, is fairly typical for Darwinian approaches to literature overall, many of which "do not really deal with literature but rather with the world behind it, and thus indeed miss the proper object of literary study." See Mellmann, "Evolutionary Psychology as a Heuristic in Literary Studies," in *The Evolution of Literature: Legacies of Darwin in European Cultures,* ed. Nicholas Saul and Simon J. James (Amsterdam: Rodopi, 2011), 304–5.
104. Hans Robert Jauss, "Literary History as a Challenge to Literary Theory," in *Norton Anthology of Theory and Criticism,* ed. Leitch et al., 1420, 1417.
105. Stanley Fish, of course, has been the most adamant voice in reader-response theory to insist that it "is interpretative communities, rather than either the text or

the reader, that produce meanings and are responsible for the emergence of formal features." See Fish, *Is There a Text in This Class? The Authority of Interpretive Communities* (Cambridge, Mass.: Harvard University Press, 1980), 14.

106. Wolfgang Iser, "Interaction between Text and Reader," in *Norton Anthology of Theory and Criticism,* ed. Leitch et al., 1524.

107. Roland Barthes, "From Work to Text," in *Norton Anthology of Theory and Criticism,* ed. Leitch et al., 1327.

108. "In a complicated way, I would hold to the statement that 'the text deconstructs itself, is self-deconstructive,' rather than being deconstructed by a philosophical intervention from outside the text." Paul de Man, "An Interview with Stefano Rosso," in *Resistance to Theory* (Minneapolis: University of Minnesota Press, 1986), 118. Derrida also does not hesitate to characterize the text as "an organism indefinitely regenerating its own tissue" after each reading. See Jacques Derrida, "From Dissemination," in *Norton Anthology of Theory and Criticism,* ed. Leitch et al., 1697.

109. Cf. Donald E. Brown, "The Universal People," in *Evolution, Literature, and Film,* ed. Boyd, Carroll, and Gottschall, 83–95; Carroll, *Reading Human Nature,* 6.

110. Ellen Spolsky, "Making 'Quite Anew': Brain Modularity and Creativity," in *Introduction to Cognitive Cultural Studies,* ed. Zunshine, 85.

111. A comprehensive and detailed overview is provided in the excellent essay by Mark J. Bruhn, "Introduction: Exchange Values: Poetics and Cognitive Science," *Poetics Today* 32, no. 3 (2011): 403–60. Bruhn distinguishes three basic approaches depending on which disciplinary perspective dominates a given research agenda: one approach has been to assimilate cultural studies into cognitive science ("cognitive science as poetics"), whereas the second approach ("poetics as cognitive science") sought to reverse that trajectory and integrate cognitive science into cultural studies. The third approach, finally, "would stress the incommensurability of humanist and scientific modes of inquiry but insist that the fullness of knowledge requires both, each operating more or less independently in its own sphere. Thus we will term this approach 'poetics and/or cognitive science'" (408). Like Bruhn, bioaesthetics is strongly committed to this last approach.

112. Stefano Franchi and Güven Güzeldere, "Of Gaps, Bridges, and Close Encounters of One of a Kind: An Introduction," *Stanford Humanities Review* 4, no. 1 (1995): http://web.stanford.edu/group/SHR/4-1/text/gaps.html.

113. See Bruhn, "Introduction," 408; Lisa Zunshine, "Part III: Cognitive Narratology," in *Introduction to Cognitive Cultural Studies,* ed. Zunshine, 53.

114. See Katja Mellmann, *Emotionalisierung—Von der Nebenstundenpoesie zum Buch als Freund: Eine emotionspsychologische Analyse der Literatur der Aufklärungsepoche* (Paderborn: Mentis, 2006); H. Porter Abbott, "Reading Intended Meaning Where None Is Intended. A Cognitivist Reappraisal of the Implied Author," *Poetics Today* 32, no. 3 (2011): 461–87; Alan Richardson, *British Romanticism and the Science of the Mind* (Cambridge, U.K.: Cambridge University Press, 2005); and Richardson, *The Neural Sublime: Cognitive Theories and Romantic Texts* (Baltimore: Johns Hopkins University Press, 2010).

115. See Alan Richardson and Ellen Spolsky, eds., *The Work of Fiction: Cognition,*

Culture, and Complexity (New York: Routledge, 2004). Others have gone further, claiming that "any investigation of cognition must start not from cognitive science's traditional assumptions, but from the form of reflection on thinking manifest in the analyses of literature and poetry." See Franchi and Güzeldere, "Of Gaps, Bridges, and Close Encounters of One of a Kind." Similarly, see Bruhn, "Introduction," 405.

116. Ellen Spolsky, "Preface," in *Work of Fiction*, ed. Richardson and Spolsky, viii.
117. The "functioning of human language depends on both its iterability *and* its instability." See Ellen Spolsky, "Darwin and Derrida: Cognitive Literary Theory as a Species of Post-Structuralism," in *Introduction to Cognitive Cultural Studies*, ed. Zunshine, 299. Similarly, "What is essential about mental imagery is at once its variability and predictability." See G. Gabrielle Starr, "Multisensory Imagery," in *Introduction to Cognitive Cultural Studies*, ed. Zunshine, 277.
118. Patrick Colm Hogan, *What Literature Teaches Us about Emotion* (Cambridge, U.K.: Cambridge University Press, 2011), 1.
119. Patrick Colm Hogan, "Literary Universals," in *Introduction to Cognitive Cultural Studies*, ed. Zunshine, 43.
120. Hogan, *What Literature Teaches Us about Emotion*, 18, 301.
121. Ibid., 8.
122. Ibid., 96, 90.
123. Ibid., 96–97, 100.
124. Lisa Zunshine, *Why We Read Fiction: Theory of Mind and the Novel* (Columbus: Ohio State University Press, 2006), 43.
125. Lisa Zunshine, "Introduction: What Is Cognitive Cultural Studies?," in *Introduction to Cognitive Cultural Studies*, ed. Zunshine, 13, 2.
126. Ibid., 15.
127. In his best seller *The Blank Slate*, Pinker famously argued that modernist (and, a fortiori, postmodern) literature unduly replaced all "the tricks that artists had used for millennia to please the human palate," such as "omniscient narration, structured plots, the orderly introduction of characters, and general readability," with "a stream of consciousness, events presented out of order, baffling characters and causal sequences, subjective and disjointed narration, and difficult prose" (409–10). In response, Zunshine rightly points out that "writers, after all, have always experimented with the palates of their readers." As pertinent examples she mentions Laurence Sterne's *Tristram Shandy* and E. T. A. Hoffmann's *Kater Murr*. She also mentions that "even the most difficult experimental modernist or postmodernist text would still have to engage the reader emotionally." See Zunshine, *Why We Read Fiction*, 172n6.
128. Zunshine, *Why We Read Fiction*, 159.
129. Spolsky, "Making 'Quite Anew,'" 87, 89.
130. Ibid., 92, 85.

5. Neuroaesthetics

1. See "This Is Your Brain on Politics," *New York Times*, November 11, 2007, "http://www.nytimes.com/2007/11/11/opinion/11freedman.html.
2. See David Biello, "Searching for God in the Brain," *Scientific American Mind*, October 2007, https://www.scientificamerican.com/article/searching-for-god-in-the-brain/.
3. Helen Fisher, *Why We Love: The Nature and Chemistry of Romantic Love* (New York: Holt, 2004), 76.
4. Cf. Changeux, *Physiology of Truth*. Some years ago, India actually admitted brain scans as evidence in legal court proceedings not only for minor cases, but in capital cases as well. See Anand Giridharadas, "India's Use of Brain Scans in Courts Dismays Critics," *New York Times*, September 15, 2008.
5. Michael S. Gazzaniga, *The Ethical Brain* (New York: Dana Press, 2005), 31.
6. Francis Crick, *The Astonishing Hypothesis: The Scientific Search for the Soul* (New York: Scribner, 1995), 3.
7. Goodman, qtd. in Regis, *What Is Life?*, 162. More specifically, it remains unclear "whether the activation indicates excitatory processes or inhibitory processes. We cannot tell whether the activation arises in a location or is an indirect consequence of another location, which releases inhibition. Moreover, it is not clear that more blood flow always occurs when there is more processing." See Stephen M. Kosslyn, *Image and Brain: The Resolution of the Imagery Debate* (Cambridge, Mass.: MIT Press, 1994), 48.
8. An electroencephalogram, for example, consists of electrodes placed on the scalp that measure electrical activity in the brain accurately in terms of time resolution (up to a thousandth of a second). However, these results do not include location specificity, meaning it is impossible to say where exactly on the cortex the measured activity is taking place. By contrast, positron emission tomography or functional magnetic resonance imaging are able to exactly locate the occurrence of neural activity, but their temporal resolution is extremely low, since they measure neural activity indirectly via the level of oxygen in the blood. So, once a particular cortical area gets activated, it usually takes several seconds before the oxygen-enriched blood arrives at its destination to support the cells' activity. In fact, some fine-tuned experiments actually show an immediate decrease in blood levels of the affected area because the neural activity quickly consumes the presently available oxygen. By the time the oxygen-rich blood then arrives at this spot, the neural activity itself may have already subsided. For a comprehensive account of fMRI technology, see Bärbel Hüsing, Lutz Jäncke, and Brigitte Tag, *Impact Assessment of Neuroimaging: Final Report* (Amsterdam: IOS Press, 2006).
9. Kosslyn, *Image and Brain*, 48.
10. Recent books on fMRI imagery and science studies have argued this point convincingly. Kelly A. Joyce provides a detailed historical analysis of MRI imagery in her book *Magnetic Appeal: MRI and the Myth of Transparency* (Ithaca, N.Y.:

Cornell University Press, 2008). She argues that MRI "images do not *reveal* the inner body, but instead *produce* the body, bringing together aspects of physical bodies, technology, and cultural, social, and economic factors in ways that both include and exceed the physical body" (48). Joseph Dumit makes a similar point: "Images are produced and selected for publication to make particular points and to illustrate the argument and other data presented, not to stand alone. They are, in other words, explicitly rhetorical. This is, one could say, the only way one can present images. . . . With each appropriation and subsequent translation, the content of the image, its qualifications and brain-type referent, changes." See Dumit, *Picturing Personhood: Brain Scans and Biomedical Identity* (Princeton, N.J.: Princeton University Press, 2004), 16. See also Michael Hagner, *Homo Cerebralis: Der Wandel vom Seelenorgan zum Gehirn* (Berlin: Berlin Verlag, 1997).

11. Cf. Roland Barthes, *Camera Lucida: Reflections on Photography*, trans. Richard Howard (New York: Hill and Wang, 1981), 76.
12. Cf. Jay David Bolter and Richard Grusin, *Remediation: Understanding New Media* (Cambridge, Mass.: MIT Press, 2000).
13. Cf. Lev Manovich, *The Language of New Media* (Cambridge, Mass.: MIT Press, 2001).
14. On the question of fallibility and scientific progress, see Peter Godfrey-Smith's excellent comments on Popper and Kuhn in his book *Theory and Reality*, 59, 92; and Michael Ruse, *Taking Darwin Seriously: A Naturalist Approach to Philosophy* (Amherst, N.Y.: Prometheus Books, 1998).
15. This is Nelson Goodman's key point in his marvelous book *Ways of Worldmaking* (Indianapolis: Hackett, 1978).
16. See Martin Skov and Oshin Vartanian, eds., *Neuroaesthetics* (New York: Baywood, 2009); Olga Pombo, Silvia Di Marco, and Marco Pina, eds., *Neuroaesthetics: Can Science Explain Art?* (Fim de Século: Sociedade Unipessoal LDA, 2010); Melissa M. Littlefield and Jenell M. Johnson, eds., *The Neuroscientific Turn: Transdisciplinarity in the Age of the Brain* (Ann Arbor: University of Michigan Press, 2012); Jon O. Lauring, ed., *An Introduction to Neuroaesthetics: The Neuroscientific Approach to Aesthetic Experience, Artistic Creativity, and Arts Appreciation* (Copenhagen: Museum Tusculanum Press, 2014); Chatterjee, *Aesthetic Brain*; and Eric Kandel, *Reductionism in Art and Brain Science: Bridging the Two Cultures* (New York: Columbia University Press, 2016). I will make reference to earlier publications in the course of this chapter.
17. "The literary critic as neuroscience groupie is part of a growing trend," Raymond Tallis lamented in the *Times Literary Supplement* back in 2008. See Tallis, "The Neuroscience Delusion," *Times Literary Supplement*, August 9, 2008. But neuroaesthetics is more than just a trend, nor is it confined to literature. John Onians, an art historian from East Anglia University who coined the term "neuroarthistory," rightly points out that the "use of neuroscientific knowledge as an aid to the study of art is now an established practice." See Onians, *Neuroarthistory: From Aristotle and Pliny to Baxandall and Zeki* (New Haven, Conn.: Yale University Press, 2008), 9. Indeed, Lauring's anthology *An Introduction to Neuroaesthetics* showcases a

broad variety of different media and artforms including literature, film, art, music, dance, et cetera, and most recent publications implicitly or explicitly refer to neuroaesthetics as a genuine *discipline* in the making. Melissa M. Littlefield and Jenell M. Johnson, for example, seek to "(re)define the field as a translational discipline" in a way similar to how I have defined bioaesthetics throughout this book. They emphasize, above all, neuroaesthetics' long and "dynamic history" dating back to the nineteenth century and beyond, as well as the fact that its methods and theories are informed by numerous other academic disciplines. Most important, they acknowledge and critically engage "the specter of reductionism, essentialism, and biologism" that continues to haunt neuroaesthetics due to its foundation in molecular biology and digital imagery. See Littlefield and Johnson, "Introduction," in *Neuroscientific Turn*, ed. Littlefield and Johnson, 3–4. Bioaesthetics commits to the same principles and pursues similar objectives as the kind of neuroaesthetics outlined by Littlefield and Johnson.

18. On the scientific side of the spectrum, for example, critics have aptly distinguished between descriptive and experimental neuroaesthetics, between empirical studies based exclusively on brain imaging techniques and those that are not, between studies that focus mainly on the *production* of art versus those that examine the *reception* of art, et cetera. On the humanities side, we should distinguish between neuronal aesthetics (like the history of brain science or neuronal media theory) and neuroarthistory, which itself is split between more empirical and more speculative approaches.
19. Cf. Onians, *Neuroarthistory*. Onians may have coined the term "neuroarthistory" in 2008, but scholars' interest in the relationship between art and the brain dates back at least to the early 1980s. Based on the proceedings of several interdisciplinary conferences that took place between 1979 and 1983, the first anthology on this subject appeared in 1988, and the editors explicitly refer to the "emerging field of neuroaesthetics" in their introduction. See Ingo Rentschler, Barbara Herzberger, and David Epstein, eds., *Beauty and the Brain: Biological Aspects of Aesthetics* (Basel: Birkhäuser, 1988), 10.
20. Onians, *Neuroarthistory,* 13.
21. Karl Clausberg, *Neuronale Kunstgeschichte: Selbstdarstellung als Gestaltungsprinzip* (New York: Springer, 1999), 9; all translations of this text are my own.
22. Onians, *Neuroarthistory,* 18, 29.
23. Zeki, *Inner Vision,* 1, 3.
24. Rafael Yuste and George M. Church, "The New Century of the Brain," *Scientific American* 310, no. 3 (2014): 38–45.
25. See Francisco Ortega and Fernando Vidal, eds., *Neurocultures: Glimpses into an Expanding Universe* (Bern: Peter Lang, 2011); David Bates and Nima Bassiri, eds., *Plasticity and Pathology: On the Formation of the Neural Subject* (New York: Fordham University Press, 2016).
26. Brain modularity has been confirmed by numerous empirical studies starting in the early twentieth century. In 1909, for example, the German neurologist Korbinian Brodmann divided the cerebral cortex into fifty-two distinct areas, assigning

each one a particular function. Though more than a century old, Brodmann's topological classification system is still being used in the field today. Brodmann areas 1, 2, and 3, for example, are located at the edge of the parietal lobe and together constitute the primary somatosensory cortex. This region of the brain is mainly responsible for processing sensory input from touch, temperature, and proprioception as well as the feeling of pain. Similarly, Brodmann area 17 is still referred to today as the primary visual cortex (also called V1 or striate cortex), even though we know that it is only one of at least thirty-two or more separate regions of the brain associated with our sense of vision.

27. The renowned historian of neuroscience Olaf Breidbach is extremely critical of the discipline due to its historical amnesia: "The theoretical foundation [of contemporary neuroscience] must be understood as a historically grown construction. The awareness of this history is . . . hence a prerequisite for explaining its own foundation to neuroscience." See Breidbach, *Die Materialisierung des Ichs: Zur Geschichte der Hirnforschung im 19. und 20. Jahrhundert* (Frankfurt: Suhrkamp, 1996), 21; my translation. Michael Hagner and Cornelius Borck agree: "Scientific representations of the mindful brain, as well as outspokenly dualistic viewpoints held by neurophysiologists like Charles S. Sherrington, were part of a particular form of culture and should be reconstructed as such." See Hagner and Borck, "Mindful Practices: On the Neurosciences in the Twentieth Century," *Science in Context* 14, no. 4 (2001): 507.
28. Ian Gold and Adina L. Roskies, "Philosophy of Neuroscience," in *The Oxford Handbook of Philosophy of Biology,* ed. Ruse, 352. The same problem, of course, exists in genomics, which also "yields an enormous list of nucleic acids that is totally devoid of meaning, and does not, by itself, give us any useful information about the genetics of the organism." See Zachery Ernst, "Philosophical Issues Arising from Genomics," in *The Oxford Handbook of Philosophy of Biology,* ed. Ruse, 312.
29. Hagner, for example, laments "the amazing theoretical poverty of today's cognitive neurosciences." See Michael Hagner, *Der Geist bei der Arbeit. Historische Untersuchungen zur Hirnforschung* (Göttingen: Wallstein, 2006), 16; my translation. Similarly, Olaf Breidbach argues that "the ideas on which neuroscience's interpretative schemes and modes of questioning are based essentially date back to an epoch that was not yet familiar with the experimental methodology of modern neuroscience." See Breidbach, "Hirn und Bewußtsein," in *Neurowissenschaften und Philosophie,* ed. Michael Pauen and Gerhard Roth (Munich: Fink, 2000), 51; my translation.
30. Christian E. Elger et al., "Das Manifest: Elf führende Neurowissenschaftler über Gegenwart und Zukunft der Hirnforschung," *Gehirn und Geist,* September 25, 2006; my translation. See http://www.gehirn-und-geist.de/artikel/852357&_z=798884.
31. Daniel C. Dennett, *Consciousness Explained* (London: Penguin, 1991), 111, 231, 113.
32. Cf. Margaret Livingstone, *Vision and Art: The Biology of Seeing* (New York: Abrams, 2002), 53.

33. Semir Zeki, *Splendors and Miseries of the Brain: Love, Creativity, and the Quest for Human Happiness* (Hoboken, N.J.: Wiley-Blackwell, 2008), 66.
34. J. Y. Lettvin, H. R. Maturana, W. S. McCulloch, and W. H. Pitts, "What the Frog's Eye Tells the Frog's Brain," in *The Mind: Biological Approaches to Its Functions,* ed. William C. Corning and Martin Balaban (New York: Interscience, 1968), 257.
35. Humberto R. Maturana, *Biologie der Realität,* trans. Wolfram Klar Köck and Gerda Verden-Zöller (Frankfurt: Suhrkamp, 1998), 20; my translation. See also Breidbach, *Deutungen,* 14ff.
36. Torsten N. Wiesel, "The Postnatal Development of the Visual Cortex and the Influence of Environment," in David H. Hubel and Torsten N. Wiesel, *Brain and Visual Perception: The Story of a 25-Year Collaboration* (Oxford, U.K.: Oxford University Press, 2005), 694.
37. Semir Zeki, *A Vision of the Brain* (Hoboken, N.J.: Wiley-Blackwell, 1993), 218, 224.
38. "Although neural plasticity is primarily evident during the maturational period, some plasticity remains in adulthood," meaning that "normal brain development is a functionally plastic epigenetic process." See Kathleen R. Gibson, "Epigenesis, Brain Plasticity, and Behavioral Versatility: Alternatives to Standard Evolutionary Psychology Models," in *Complexities: Beyond Nature and Nurture,* ed. Susan McKinnon and Sydel Silverman (Chicago: University of Chicago Press, 2005), 31–32. Gerald M. Edelman agrees: "Indeed, even after the major outlines of adult neuroanatomy are established, the borders of cortical maps can change dynamically, depending on the input from the body and the environment." See Edelman, *Second Nature: Brain Science and Human Knowledge* (New Haven, Conn.: Yale University Press, 2006), 56. Similarly, see Colin Blakemore, "How the Environment Helps to Build the Brain," in *Mind, Brain, and the Environment,* ed. Bryan Cartledge (Oxford, U.K.: Oxford University Press, 1998), 28–56; and Vernon B. Mountcastle, "Preface to the Princeton Edition," in Jean-Pierre Changeux, *Neuronal Man: The Biology of Mind,* trans. Laurence Garey (Princeton, N.J.: Princeton University Press, 1985), xiii–xv.
39. Gerald M. Edelman, "Building a Picture of the Brain," in *The Brain,* ed. Gerald M. Edelman and Jean-Pierre Changeux (New Brunswick, N.J.: Transaction, 2001), 33.
40. "It is specificity and plasticity rather than nature and nurture that provide the dialectic within which development occurs, and both are utterly dependent on both genes and environment." See Rose, *Future of the Brain,* 64.
41. Kandel, *Reductionism in Art and Brain Science,* 52. See also *Eric Kandel: In Search of Memory,* directed by Petra Seeger (Icarus Film, 2008).
42. Changeux, *Physiology of Truth,* 62.
43. Edelman, *Second Nature,* 27.
44. Joseph LeDoux, *Synaptic Self: How Our Brains Become Who We Are* (New York: Penguin, 2003), 79, 77, 307.
45. Margulis and Sagan, *What Is Life?,* 150, 233.
46. "There are many other understandings and uses of the term consciousness," Steven Rose points out. "There is Freudian consciousness, with its murky unconscious

world of desires and fears. There is social consciousness, class, ethnic, or feminist consciousness, the recognition of having a standpoint from which one can interpret and act upon the world. All these understandings are lost in the impoverished worlds shared by these philosophers and neuroscientists who reduce such varied worlds to that of being aware, being awake, being unanaesthetised. Being conscious is more than this." See Rose, "Escaping Evolutionary Psychology," in *Alas, Poor Darwin,* ed. Rose and Rose, 304.

47. Dennett, *Consciousness Explained,* 448.
48. Hence Antonio Damasio distinguishes between proto self, core self, and autobiographical self, while John G. Taylor refers to what he calls passive (or perceptual) self, active self, and self-consciousness. See Damasio, *Self Comes to Mind: Constructing the Conscious Brain* (New York: Vintage Books, 2012), 22ff.; and Taylor, *The Race for Consciousness* (Cambridge, Mass.: MIT Press, 1999), 30. Using a binary model, Joseph LeDoux, by contrast, only distinguishes between "implicit" and "explicit self" depending on whether relevant information is stored in implicit or explicit memory. See LeDoux, *Synaptic Self,* 26ff.
49. Patricia and Paul Churchland, qtd. in Susan Blackmore, *Conversations on Consciousness: What the Best Minds Think about the Brain, Free Will, and What It Means to Be Human* (Oxford, U.K.: Oxford University Press, 2007), 55.
50. Crick, *Astonishing Hypothesis,* 3.
51. Chalmers, qtd. in Blackmore, *Conversations on Consciousness,* 38.
52. Blackmore, *Conversations on Consciousness,* 5.
53. Thomas Nagel, "What Is It Like to Be a Bat?," *Philosophical Review* 83, no. 4 (1974): 435–50; John R. Searle, "Mind, Brains, and Programs," *Behavioral and Brain Sciences* 3, no. 3 (1980): 417–57; David J. Chalmers, *The Conscious Mind: In Search of a Fundamental Theory* (New York: Oxford University Press, 1996).
54. Thompson, *Mind in Life,* 164.
55. Jean-Pierre Changeux and Paul Ricoeur, *What Makes Us Think? A Neuroscientist and a Philosopher Argue about Ethics, Human Nature, and the Brain,* trans. M. B. DeBevoise (Princeton, N.J.: Princeton University Press, 2002), 47.
56. Werner Heisenberg, *Physics and Philosophy: The Revolution in Modern Science* (New York: Harper Perennial, 2007), 148.
57. Alain Badiou, *Being and Event,* trans. Oliver Feltham (London: Bloomsbury, 2013). Badiou's claim has been sharply criticized by some mathematicians. See Ricardo L. Nirenberg and David Nirenberg, "Badiou's Number: A Critique of Mathematics as Ontology," *Critical Inquiry* 37, no. 4 (2011): 583–614.
58. Edmund Husserl, *The Crisis of European Sciences and Transcendental Phenomenology,* trans. David Carr (Evanston, Ill.: Northwestern University Press, 1970), 92.
59. Rorty, *Truth and Progress,* 90.
60. Bazon Brock, "Neuronale Ästhetik," 1996, http://www.brock.uni-wuppertal.de/Projekte/NeuroAe1.html; all English translations of this text are my own.
61. See Gadamer, *Truth and Method.*
62. Olaf Breidbach, "The Beauties and the Beautiful: Some Considerations from the

Perspective of Neuronal Aesthetics," in *Evolutionary Aesthetics,* ed. Eckart Voland and Karl Grammer (Berlin: Springer, 2003), 40, 65, 56, 62.

63. Barbara Maria Stafford, *Echo Objects: The Cognitive Work of Images* (Chicago: University of Chicago Press, 2009), 2.
64. Barbara Maria Stafford, "Hedonics," in *Sensorium: Embodied Experience, Technology, and Contemporary Art,* ed. Caroline Jones (Cambridge, Mass.: MIT Press, 2006), 152.
65. Stafford, *Echo Objects,* 100.
66. Stafford, "Hedonics," 152.
67. Stafford, *Echo Objects,* 45.
68. Ibid., 154.
69. Barbara Maria Stafford, "Romantic Systematics and the Genealogy of Thought: The Formal Roots of a Cognitive History of Images," *Configurations* 12, no. 3 (2004): 317. Similarly, see Stafford, *Echo Objects,* 136.
70. Stafford, *Echo Objects,* 142–43, 148.
71. Ibid., 3.
72. Warren Neidich, *Blow-Up: Photography, Cinema, and the Brain* (Riverside: California Museum of Photography, 2003), 21.
73. Ibid., 35.
74. Ibid., 137, 163.
75. Ibid. Lutz Koepnick makes a similar argument in his excellent book *On Slowness: Toward an Aesthetics of the Contemporary* (New York: Columbia University Press, 2014).
76. Friedrich Kittler, "There Is No Software," *C-Theory: Theory, Technology, Culture* 32 (October 1995): http://www.f.waseda.jp/sidoli/Kittler_1992_No_Software.pdf.
77. Friedrich Kittler, "On the Implementation of Knowledge: Toward a Theory of Hardware," n.d., http://hydra.humanities.uci.edu/kittler/implement.html.
78. Martin Skov, "Neuroaesthetic Problems: A Framework for Neuroaesthetic Research," in *Neuroaesthetics,* ed. Skov and Vartanian, 11.
79. Martin Skov and Oshin Vartanian, "Introduction: What Is Neuroaesthetics?," in *Neuroaesthetics,* ed. Skov and Vartanian, 3.
80. Chiara Cappelletto makes a similar point, namely, that neuroaesthetics "introduces a hyperaesthetic conception of art, according to which art = aesthetics = visual brain. Artist and spectator as complementary participants of the aesthetic experience would then disappear." See Cappelletto, "Neuroaesthetics: Origins, Perspectives, Problems," in *Neuroaesthetics,* ed. Pombo, Di Marco, and Pina, 87.
81. Jean-Pierre Changeux, "Art and Neuroscience," *Leonardo* 27, no. 3 (1994): 190.
82. Livingstone, *Vision and Art,* 37–40.
83. Gitte Orskou, *Olafur Eliasson: The Blind Pavilion* (Copenhagen: Danish Contemporary Art Foundation, 2003), 21.
84. Alva Noe, "Experience and Experiment in Art," *Journal of Consciousness Studies* 7, nos. 8–9 (2000): 128.
85. Chatterjee, *Aesthetic Brain,* 141.

86. Ibid., 183.

87. Skov and Vartanian, "Introduction," 4. Brown and Dissanayake agree: "Strictly speaking, it is this broad area—not works of art alone—that defines the domain of neuroaesthetics," they concede. Hence, a "better goal for the field would be to develop a general, superordinate theory of aesthetic responses that applies to all appraised objects rather than one linked to artworks." While this reorientation would certainly improve the conceptual coherence of the field, it would also highlight the fact that neuroaesthetics has little to tell us about art and its history. See Steven Brown and Ellen Dissanayake, "The Arts Are More Than Aesthetics: Neuroaesthetics as Narrow Aesthetics," in *Neuroaesthetics,* ed. Skov and Vartanian, 44–45. See also Chatterjee, *Aesthetic Brain,* 115.

88. Skov and Vartanian, "Introduction," 6. More references to Fechner can be found in Marcos Nadal, Antoni Gomila, and Alejandro Gàlvez-Pol, "A History for Neuroaesthetics," in *Introduction to Neuroaesthetics,* ed. Lauring, 10–17; and Chatterjee, *Aesthetic Brain,* 134.

89. Cf. Dieter Kliche, "Ästhetik des Schönen, Ästhetik des Häßlichen: Akademisierung und Neuansätze im 19. Jahrhundert," in *Grundbegriffe der Ästhetik,* ed. Barck et al., 1:375.

90. V. S. Ramachandran and William Hirstein, "The Science of Art: A Neurological Theory of Aesthetic Experience," *Journal of Consciousness Studies* 6, nos. 6–7 (1999): 17.

91. Wilson, *Consilience,* 231.

92. Ramachandran and Hirstein, "Science of Art," 15–18.

93. Here are two typical comments from scientists: "I applaud the efforts of Ramachandran and Hirstein to foster dialogue among artists, visual physiologists, and evolutionary biologists." Amy Ione, "Connecting the Cerebral Cortex with the Artists' Eyes, Mind, and Culture," *Journal of Consciousness Studies* 7, nos. 8–9 (2000): 26. "R&H have written a vigorous, thought provoking paper." Bruce Mangan, "It Don't Mean a Thing If It Ain't Got That Swing," *Journal of Consciousness Studies* 6, nos. 6–7 (1999): 56. Here are two responses from scholars in the humanities: "I find it hard to take their argument seriously," one reviewer wrote, because it fails to "respect the rigorous debates about aesthetic experience that are part of differing academic disciplines and cultural traditions." Ruth Wallen, "Response to Ramachandran and Hirstein," *Journal of Consciousness Studies* 6, nos. 6–7 (1999): 69, 72. "Our response to art or aesthetics may have some biological elements," another critic concedes, "but it is culture that provides the unique qualities of an artistic tradition." Partha Mitter, "A Short Commentary on 'The Science of Art,'" *Journal of Consciousness Studies* 6, nos. 6–7 (1999): 65.

94. E. H. Gombrich, "Concerning 'The Science of Art': Commentary on Ramachandran and Hirstein," *Journal of Consciousness Studies* 7, nos. 8–9 (2000): 17.

95. As Mitter points out in her response to Ramachandran and Hirstein, Gombrich "only spoke of the laws of perception that affected representation but did not claim that such perception necessarily led to aesthetic pleasure. In other words,

he would, I am sure, distinguish between our cognitive ability and our propensity for pleasure. It is this eliding of cognition and pleasure in Ramachandran & Hirstein that I find rather unconvincing." See Mitter, "A Short Commentary on 'The Science of Art,'" 64.

96. Donnya Wheelwell, "Against the Reduction of Art to Galvanic Skin Response," *Journal of Consciousness Studies* 7, nos. 8–9 (2000): 42.
97. In a 2004 study cowritten with his colleague Hideaki Kawabata, Zeki explicitly excludes from consideration "the important question of how what an individual regards as beautiful is conditioned by culture, upbringing, and inclination. While acknowledging these important issues, we have tried to circumvent them by allowing subjects to determine themselves what is beautiful and what is not." Hideaki Kawabata and Semir Zeki, "Neural Correlates of Beauty," in *Neuroaesthetics*, ed. Pombo, Di Marco, and Pina, 52.
98. Zeki, *Inner Vision*, 9–10; italics removed.
99. Ibid., 42, 26.
100. Ibid., 29.
101. Zeki, *Vision of the Brain*, 202.
102. Zeki, *Inner Vision*, 121–22.
103. Ibid., 119.
104. Ibid., 120.
105. Ibid., 120, 124.
106. This, indeed, is Clausberg's claim in his *Neuronale Kunstgeschichte*, 22.
107. Zeki, *Inner Vision*, 113.
108. Ibid., 120.
109. Ibid., 124.
110. "It is neurologically impossible to conceive visually of ideal forms without a brain that has been exposed to the visual world from birth," meaning "that there are no ideal forms that have an existence in the outside world without reference to the brain." Ibid.,40.
111. Cf. Althusser, "Philosophy and the Spontaneous Philosophy of the Scientists."
112. Zeki, *Inner Vision*, 45.
113. Ibid., 37; John Dewey, *Art as Experience* (New York: Perigee, 2005), 95.
114. Cf. Deleuze, *Difference and Repetition*, 189.
115. Gilles Deleuze, *The Logic of Sense*, ed. Constantin V. Boundas, trans. Mark Lester (New York: Columbia University Press, 1990), 262.
116. Gilles Deleuze, *Negotiations*, 126.
117. Deleuze, *Difference and Repetition*, 56; my emphasis.
118. Ibid., 56–57.

Coda

1. Gilles Deleuze and Félix Guattari, *A Thousand Plateaus: Capitalism and Schizophrenia*, trans. Brian Massumi (Minneapolis: University of Minnesota Press, 1987), 158.

2. Gilles Deleuze, *Spinoza: Practical Philosophy,* trans. Robert Hurley (San Francisco: City Light Books, 1988), 218, 226. Similarly, see Brian Massumi, "Notes on the Translation," in Deleuze and Guattari, *Thousand Plateaus,* xvi.
3. Gilles Deleuze and Félix Guattari, *What Is Philosophy?* trans. Hugh Tomlinson (New York: Columbia University Press, 1990), 169.
4. Affect, Patricia Ticineto Clough agrees, draws us "to and beyond a techno-ontological threshold" and asks us "to think preindividual affectivity, where the politics of mourning ghosted bodies is too slow to engage preindividual affective capacities that a capitalist political economy already maps and mines." See Clough, "Introduction," in *The Affective Turn: Theorizing the Social,* ed. Patricia Ticineto Clough and Jean Halley (Durham, N.C.: Duke University Press, 2007), 28.
5. Hence Brian Massumi refers to the "primacy of the affective," and Marco Abel speaks of the "priority of affect over cognition." See Massumi, *Parables of the Virtual: Movement, Affect, Sensation* (Durham, N.C.: Duke University Press, 2002), 24; Abel, "Intensifying Affect," *Electronic Book Review,* October 24, 2008, 4. Most essays in Clough and Halley's collection *The Affective Turn* also refer to Deleuze in juxtaposing affect to cognition. Similarly, Kathleen Stewart describes affects as "*something*" that "work[s] not through 'meanings' per se." See Stewart, *Ordinary Affects* (Durham, N.C.: Duke University Press, 2007), 3–4. Simon O'Sullivan agrees: "Affects are . . . the stuff that goes on beneath, beyond, even parallel to signification. . . . You cannot read affects, you can only experience them." See O'Sullivan, "The Aesthetics of Affect: Thinking Art beyond Representation," *Angelaki* 6, no. 1 (2001): 126. For a well-argued critique of this view, see Teresa Brennan's *The Transmission of Affect* (Ithaca, N.Y.: Cornell University Press, 2004); and Clare Hemmings's essay "Invoking Affect," in *Cultural Studies* 19, no. 5 (2005): 548–67. Hemmings acknowledges the importance of affect for contemporary studies of culture, but rejects "the contemporary fascination with affect as outside social meaning." Instead, she insists that "affect might in fact be valuable precisely to the extent that it is not autonomous" (565). I concur.
6. Jaak Panksepp, *Affective Neuroscience: The Foundations of Human and Animal Emotions* (Oxford, U.K.: Oxford University Press, 1998), 301. Again and again, Panksepp emphasizes "how deceptively wrong our explanatory schemes may be if we do not take the nature of emotional consciousness into consideration" (34). Similar to Panksepp, Joseph LeDoux also cautions against "overemphasiz[ing] the contribution of cognitive processes in emotion." Yet he, too, acknowledges that "emotion involves action tendencies and bodily responses, as well as conscious experiences." See LeDoux, *The Emotional Brain: The Mysterious Underpinnings of Emotional Life* (New York: Simon and Schuster, 1998), 52. Antonio Damasio agrees. At the most basic level, he argues, emotions comprise "relatively simply, stereotyped patterns of response which include metabolic regulation, reflexes," and other unconscious processes. Yet even though "a state of emotion . . . can be triggered and executed nonconsciously," there nonetheless exist numerous "bridges between emotion and consciousness," meaning that "emotion is integral to the processes of reasoning and decision making." See Damasio, *The Feeling of What*

Happens: Body and Emotion in the Making of Consciousness (San Diego: Harcourt, 2000), 55, 37, 41.

7. Rose, *Future of the Brain,* 103.
8. Ruth Leys convincingly made this case in "The Turn to Affect: A Critique," in *Critical Inquiry* 37, no. 3 (2011): 434–72.
9. Melissa Gregg and Gregory J. Seigworth, "An Inventory of Shimmers," in *The Affect Theory Reader,* ed. Gregg and Seigworth (Durham, N.C.: Duke University Press, 2010), 14.
10. Cf. Carsten Strathausen, "Adorno; or, The End of Aesthetics," in *Globalizing Critical Theory,* ed. Max Pensky (New York: Rowman & Littlefield, 2005), 221–40.
11. William K. Wimsatt Jr. and Monroe C. Beardsley, "The Affective Fallacy," in *Norton Anthology of Theory and Criticism,* ed. Leitch et al., 1246.
12. Antonio Damasio, for example, uses the term "emotion" to refer to people's unconscious bodily reactions to internal or external stimuli. Although more developed than affects, emotions, for Damasio, are still "part of the basic mechanism of life regulation." Feelings, on the other hand, signify higher cognitive processes and "mental events" in the neocortex. In short, "emotions precede feelings," and affects precede emotions. See Damasio, *Looking for Spinoza: Joy, Sorrow, and the Feeling Brain* (London: Harvest, 2003), 28–29. Teresa Brennan agrees: "Feelings [are] sensations that have found the right match in words" (*Transmission of Affect,* 5). Yet unlike Damasio, Brennan skips the term "emotion" and juxtaposes feelings directly to affects. Brian Massumi, finally, does the exact opposite: he juxtaposes affects to emotions but rarely ever mentions the term "feeling" *(Parables of the Virtual).* Given the semantic ambiguity of their basic concepts, it is often difficult to read these critics together without getting confused about what, exactly, they mean to say.
13. Nigel Thrift distinguishes four different approaches to affect: the phenomenological tradition (including social interactionism and hermeneutics); the psychoanalytical framework centered around drives; the naturalist approach of "a world which is constantly becoming" (i.e., Deleuze); and Darwinian evolutionary theory (including Silvan Tomlins's influential theory of affect). Cf. Thrift, "Intensities of Feeling: Towards a Spatial Politics of Affect," *Geografiska Annaler* 86, no. 1 (2004): 57–78. Eve Kosofsky Sedgwick's intriguing study of affect, for example, is based on Tomlins's rather than Deleuze's theory. See Sedgwick, *Touching Feeling: Affect, Pedagogy, Performativity* (Durham, N.C.: Duke University Press, 2003).
14. Both Teresa Brennan and Jaak Panksepp provide an excellent overview not only of the various terms employed today, but also of their historical usage and change of meaning since the eighteenth century. See Brennan, *Transmission of Affect,* 5–6, 101–15; and Panksepp, *Affective Neuroscience.*
15. Massumi, *Parables of the Virtual,* 28, 260, 36.
16. See John Protevi, *Political Affect: Connecting the Social and the Somatic* (Minneapolis: University of Minnesota Press, 2009); and Protevi, *Life, War, Earth.*
17. Lauren Berlant, *Cruel Optimism* (Durham, N.C.: Duke University Press, 2011).
18. Rosi Braidotti, *The Posthuman* (Oxford, U.K.: Polity Press, 2013), 2, 49. The

Deleuzian philosopher and political theorist William Connolly likewise insists that we still live "in a world in which humanity matters immensely," and he explicitly mentions Kant "as a prescient genius to be engaged" more seriously by posthumanist theory. See Connolly, *The Fragility of Things: Self-Organizing Processes, Neoliberal Fantasies, and Democratic Activism* (Durham, N.C.: Duke University Press, 2013), 50, 17.

19. Deleuze and Guattari, *What Is Philosophy?,* 198.
20. Gilles Deleuze, *Cinema II: The Time Image*, trans. Hugh Tomlinson and Robert Galeta (Minneapolis: University of Minnesota Press, 1989), 205.
21. Deleuze, *Logic of Sense,* 52.
22. Gilles Deleuze, *Essays Critical and Clinical,* trans. Michael A. Greco and Daniel W. Smith (Minneapolis: University of Minnesota Press, 1997), 90, 4.
23. Deleuze and Guattari, *What Is Philosophy?,* 202–3.
24. Gilles Deleuze, *Francis Bacon: The Logic of Sensation,* trans. Daniel W. Smith (Minneapolis: University of Minnesota Press, 2002), 39.
25. Gilles Deleuze, *Nietzsche and Philosophy,* trans. Hugh Tomlinson (New York: Columbia University Press, 1983), 28. "Deleuze's aesthetic theory," Daniel W. Smith agrees, "is not a theory of reception, an analytic of the spectator's judgment of a work of art, but a theory of aesthetics written from the point of view of creation." See Smith, *Essays on Deleuze* (Edinburgh: Edinburgh University Press, 2012), 104.
26. Gilles Deleuze, *Desert Islands and Other Texts,* ed. David Lapoujade, trans. Mike Taormina (Cambridge, Mass.: MIT Press, 2004) 93.
27. Deleuze and Guattari, *What Is Philosophy?,* 212.
28. Félix Guattari, *Chaosmosis: An Ethico-Aesthetic Paradigm,* trans. Paul Bains and Julian Pefanis (Sydney: Power Publications, 2012), 9, 93.
29. Deleuze, *Negotiations,* 143.
30. Deleuze and Guattari, *Thousand Plateaus,* 240.
31. Deleuze, *Francis Bacon,* 45. Echoing Deleuze, Grosz, too, insists that Bacon paints sensations: "The scream [in Bacon's pope paintings] only functions as sensation to the extent that we can feel and hear it, that it vibrates as a scream, that is, as visual, it nevertheless functions as an auditory cry, resonating or vibrating through us as a scream." See Grosz, *Chaos, Territory, Art: Deleuze and the Framing of the Earth* (New York: Columbia University Press, 2008), 81.
32. "Aesthetic reflection about affects [in the eighteenth century] presupposes a specific understanding of both the nature of human affects and their ideal conditioning to serve practical action and social behavior." Barck et al., eds., *Ästhetische Grundbegriffe,* 1:17; my translation.
33. "If freedom means anything for Deleuze it isn't a matter of human liberty but of liberation *from* the human." See Peter Hallward, *Out of This World: Deleuze and the Philosophy of Creation* (London: Verso, 2006), 139.
34. Protevi, *Life, War, Earth,* 156, 182.
35. Steven Shaviro, *The Universe of Things: On Speculative Realism* (Minneapolis: University of Minnesota Press, 2014), 153.

36. Jacques Derrida, *Aporias,* trans. Thomas Dutoit (Stanford, Calif.: Stanford University Press, 1993), 22.
37. Jacques Derrida, "Intellectual Courage: An Interview by Thomas Assheuer," in *Culture Machine* 2 (2000): http://www.culturemachine.net/index.php/cm/article/viewArticle/303/288.
38. George Spencer-Brown, *Laws of Form* (New York: Bantam, 1973), v.
39. Lakoff and Johnson, *Philosophy in the Flesh,* 17.
40. Canguilhem, *Knowledge of Life,* 70.

INDEX

(continued from page ii)

23 Vampyroteuthis Infernalis: A Treatise, with a Report by the Institut Scientifique de Recherche Paranaturaliste
Vilém Flusser and Louis Bec

22 Body Drift: Butler, Hayles, Haraway
Arthur Kroker

21 HumAnimal: Race, Law, Language
Kalpana Rahita Seshadri

20 Alien Phenomenology, or What It's Like to Be a Thing
Ian Bogost

19 CIFERAE: A Bestiary in Five Fingers
Tom Tyler

18 Improper Life: Technology and Biopolitics from Heidegger to Agamben
Timothy C. Campbell

17 Surface Encounters: Thinking with Animals and Art
Ron Broglio

16 Against Ecological Sovereignty: Ethics, Biopolitics, and Saving the Natural World
Mick Smith

15 Animal Stories: Narrating across Species Lines
Susan McHugh

14 Human Error: Species-Being and Media Machines
Dominic Pettman

13 Junkware
Thierry Bardini

12 A Foray into the Worlds of Animals and Humans, *with* A Theory of Meaning
Jakob von Uexküll

11 Insect Media: An Archaeology of Animals and Technology
Jussi Parikka

10 Cosmopolitics II
Isabelle Stengers

9 Cosmopolitics I
Isabelle Stengers

8 What Is Posthumanism?
Cary Wolfe

7 Political Affect: Connecting the Social and the Somatic
John Protevi

6 Animal Capital: Rendering Life in Biopolitical Times
Nicole Shukin

5 Dorsality: Thinking Back through Technology and Politics
David Wills

4 Bíos: Biopolitics and Philosophy
Roberto Esposito

3 When Species Meet
Donna J. Haraway

2 The Poetics of DNA
Judith Roof

1 The Parasite
Michel Serres

CARSTEN STRATHAUSEN is Catherine Paine Middlebush Chair of Humanities and professor of German and English at the University of Missouri. He is author of *The Look of Things: Poetry and Vision around 1900,* the editor of *A Leftist Ontology* (Minnesota, 2009), and the translator of Boris Groys's *Under Suspicion.*

CPSIA information can be obtained
at www.ICGtesting.com
Printed in the USA
LVOW03s0054200917
549322LV00003B/3/P

9 781517 900755